OFFICERS' MESS

Being a history of mess origins and customs
from a wealth of military records;
enlivened anecdotes of mess times remembered
from a host of one-time mess members;
and the progress of Charles Oswald Littlewart
from 2/Lieutenant to Major-General.

COMPILED BY LT. COL. R. J. DICKINSON

Published by

m

MIDAS BOOKS
12 DENE WAY · SPELDHURST · TUNBRIDGE WELLS · KENT

Compiled by Lt. Col. 'Dickie' Dickinson (Essex Regt. and RAOC)
Research by Ian Morley-Clarke (Parachute Regt. & Queen's)
Illustrated by Lt. Col. Frank Wilson (Royal Scots Fusiliers and 19th King George V's Own Lancers IA)
Foreword by Major General Sir 'Gerry' Duke, KBE, CB, DSO, DL (late RE)

ISBN 0 85936 094 6

Revised Edition 1977

Produced by PJH Graphic Services
Printed in England by Chapel River Press, Andover, Hants.

Contents

Illustrations **Page**

Foreword

It has been said that at any one point of time the way of life in the Army reflects the habits and customs of society as a whole, and this is perhaps particularly true of the officers' mess. It could also be said that, as with service pay, changes in social habits in the Army, which tends to be conservative, lag behind the corresponding changes in civilian life. However, the fact is that social customs in this country have changed dramatically since the last world war, and are still changing, and a way of life familiar to the older generation in the days of their youth now seems to belong to another world. This pattern has been reflected in the officers' mess, which has also seen tremendous changes in recent years. For example, the young officer of today would be pretty upset if he was told he had to wear mess-kit for dinner every evening, whereas forty years ago it was taken as a matter of course.

A British characteristic, sometimes misunderstood by our friends, and always by our enemies, is our tendency to joke about what should be serious subjects. In fact this characteristic can be a source of great strength, as has often been shown by the capacity of the British soldier to produce a light-hearted remark when confronted with some situation fraught with danger and possible disaster.

This account of officers' mess life and customs, then, is written in a light-hearted manner, but a light-heartedness not to be confused with frivolity. The editors have attempted to recapture the spirit of those days when life was lived at a more leisurely pace. They have gone to great lengths to discover and record the habits and customs of various regiments, many of them dating back to the earliest days of the Regular Army. Some of these customs have disappeared through changes in mess life, others have been lost through the amalgamation of regiments; though happily many still survive, sometimes in modified form. By gathering together in one volume this record of mess life and customs over a period which has now gone forever the editors have produced not only a book which should be fun to read for a very wide audience, but also a book which is a valuable addition in its own right to military history. I can strongly commend it.

Sir Gerald Duke

Preface

Clanger Littlewart does not exist. Or rather, his name never appeared in any Army List, but the officers who have contributed to create him most certainly did, so that he may not unjustly claim to have had an existence. Nearly everything that is ascribed to him did happen to some officer, somewhere, at some time. And the escapades, or something like them, will doubtless continue with succeeding generations of young officers. The Littlewarts of the army are immortal. Are there not a little of his hopes and fears, his successes and failures, the bricks or the bouquets, in all of us from time to time?

Of one thing we can be certain; the Ramrods are fictitious but they are real to the extent that they reflect in some measure the service of Littlewart and perhaps the gaiety and when needed the determination of the British infantry and those officers and men who have contrived to add lustre to its name.

To the reader who complains that the book has too light hearted an approach I can only say that in the days recalled, the approach to life by army officers was light hearted – though often with little enough cause. Without benefit of oxygen supply, patent foods, lightweight tents and armies of sherpas, men attacked Mount Everest wearing plus-four suits, trilby hats and no more equipment than a stout alpenstock – and got within a few thousand feet of the summit, thinking nothing of the feat. Light hearted indeed, and surely displaying a true British characteristic, particularly in times of the utmost stress. Littlewart, like his true-life contemporaries, was thus nothing if not light hearted, but he became dedicated as well.

Every effort has been made to ensure that details are as accurate as possible, but mistakes are an unfortunate fact of life, and errors may have crept in. It is hoped that they may only be small ones and not clangers for, as Littlewart himself discovered, a dropped stitch can be picked up, but a dropped brick – never!

Acknowledgements

A book of this nature cannot be compiled without the help of contributors, and I am most sincerely grateful to all those officers, serving and retired, who co-operated by providing such interesting material of every description, from personal anecdotes to Standing Orders! It was all most valuable. One contributor captured the spirit of the project to the extent of reminding me of a regimental custom of the Ramrods dating from the American War of Independence. It has completely slipped my memory and though not included here it will have an honoured place in the history of that illustrious regiment when it comes to be written.

I would like to thank the Colonels of all those regiments who have helped with contributions. My special thanks are also due to:- Lt-Gen. Sir William Oliver, GBE, KCB, KCMG, DL; Maj-Gen. Sir Gerald Duke, KBE, CB, DSO, DL; Maj-Gen. C.W. Dunbar, CBE; Maj-Gen. P. Gleadell, CB, CBE, DSO; Maj-Gen. R.F.K. Goldsmith, CB, CBE; Maj-Gen. J.D. Lunt, CBE; Brigadier C.F.O. Breese, CBE; Brigadier G.A. Rimbault, CBE, DSO, MC, DL; Brigadier W.D. Tarr; Colonel I.B. Baillie; Colonel C.R. Buchanan; Colonel J.M. Forbes, DL, JP; Colonel H.R. Grace, OBE, DL, JP; Colonel E.D. Harding, DSO; Colonel P.S. Newton, MBE; Lt-Col. R.G. Coates; Lt-Col. G.P. Gofton-Salmond, OBE; Lt-Col. A. Joanny, MBE; Lt-Col. D.A. Johnson; Lt-Col. R.D. MacLagan, CBE, MC; Lt-Col. G. Malcolm of Poltalloch, DL, JP; Col. A.G.P. Monro; Lt-Col. C.R. Murray Brown, DSO; Lt-Col. R.F. Nixon, MBE; Lt-Col. J.D. Ricketts, DSO, JP; Lt-Col. C.L. Speers; Lt-Col. G.N.B. Spencer; Lt-Col. D.V.W. Wakely, MC; Lt-Col. R.D. Wardleworth, MC; Lt-Col. J.K. Windeatt, OBE; Major J.R. Arthur; Major B.W.R. Bajer; Major M.H. Bates; Major D. Baxter; Major J.B.B. Cockcroft; Major G.M.B. Colenso-Jones; Major R.M. Collins; Major F.J. Hawley; Major D.Y. Hext; Major E. Jessup; Major E.L. Kirby, MC, TD, DL; Major J. Onslow; Major F.J. Reed; Major D.C. Risby; Major J.W.F. Rucker; Major J.P. Smiley; Major T.R. Stead, DL; Major P.W. Symonds; Major J.M.A. Tamplin; Major D.T. Tewkesbury, MBE, DL; Major W. Urban-Smith; Captain C.J.T. Davis; Mr. C.A. Fox; Mr. T. Nelson; Mr. A.A. Shipton; Mr. Maldwyn Williams; Mr. S.A. Colley.

I am also most grateful for the splendid contributions, pictures and photographs made available, and the excellent drawings by Lt.-Colonel Frank Wilson. They speak for themselves and eloquently illustrate the atmosphere of the officers' mess, past and present. Also for transparencies and photographs produced by Michael Clarke.

Finally, a large debt of gratitude is owed to the office of the Director of Public Relations who were most helpful in providing material and information and so generous with encouragement at the outset of the project, and to the Army Museums' Ogilby Trust and the National Army Museum.

To all those officers who made the compilation of this book such an interesting and worthwhile task, I offer my grateful thanks.

Officers' Mess editorial team photographed together preceding the Army v Navy Hockey match in 1947. (Ranks of 1947) Brigadier 'Gerry' Duke seated left, Lt. Col. 'Dickie' Dickinson (seated right) and Lt. Ian Morley-Clarke (standing).

Lt. Col. 'Frankie' Wilson - Queen's Messenger on the Great Wall of China. Peking c 1955.

Many officers, both serving and retired, may have anecdotes or tales of entertaining mess happenings which they feel are worthy of inclusion in any subsequent edition of this book. The publishers would welcome such contributions.

i

Officers' Mess

'Officers' mess? There ain't no officers' mess,
you just mucks in where you can.'

Such was the reply given by an unknown soldier in the Transit Camp at Southampton in September 1939 to the enquiry 'Where's the officers' mess?' made by a group of officers, the advance party of their HQ, searching for accommodation after dark.

Correct or incorrect, the answer found no favour with the senior officer of the party and it followed with inevitable army logic that the junior officer should do something about it. So Littlewart found himself ordered to 'ring up and arrange accommodation in a hotel'. An excellent idea, obviously shared by others, for Littlewart found when at last he got into the telephone booth that the appropriate section in the local classified directory had been torn out. However, recollections of a shady weekend back in the past stood him in good stead and he was able to call the Polygon Hotel, so that in a short while the officers were enjoying a meal there.

It might thus be claimed that their officers' mess was established after the custom of nearly a couple of hundred years previously, when the mess was no more than a room in a favourite tavern where the officers congregated for the midday meal.

The word 'mess', of course, originally meant no more than a portion of food, a mess of pottage. Later it came to mean the taking of meals together and, at sea, the place where they were taken – the mess decks. In the early days, however, in the army the mess meant little more than the meal (the phrase 'at mess' is still used as being at the dinner table). The officers lived in lodgings and met only for a friendly and hearty lunch. Reports exist of a certain high personage who dined with the officers 'at mess' and never exceeded his ration of half a pint of port. It is unlikely that there was an afternoon parade – for him at any rate.

The Royal Artillery mess at Woolwich probably has the distinction of being the first purpose-built mess in the present sense of the word. A mess was included in the building of the barracks in 1783, although in the field the idea of a mess was of even earlier date. In the *Military Guide for Young Officers* published in 1772 were laid down the principles of establishing a mess in the field. 'Each field officer and captain is to contribute six guineas and each subaltern and staff officer one day's pay each to the purchasing of a dining tent, a cart and two horses, table linen and kitchen furniture.' Clearly Field-Marshal Montgomery was not the first to appreciate the importance of a proper officers' mess in the field.

It is no achievement to live in squalor on active service. It is merely good sense to try to live as comfortably as the circumstances permit and ingenuity can contrive. Of the elephants that Hannibal coaxed over the Alps one can be sure that many were loaded with fellow comforts if not bed-fellow comforts. On the other hand, the luxury of the lunch at Barwa during a Frontier expedition, illustrated in the sketch, is perhaps over-doing it! But enjoy the smooth when it is available, there is rough enough!

Officers' messes have travelled a long road since 1783. What were they like in the twenties and thirties? The reminiscences of a member[1] describe life as it was then.

1. Major General R.F.K. Goldsmith late DCLI.

Lunch at Barwa, September 9th, 1897. *See page 116.*
Umra Khan's brother, Mahomed Shah Khan, sends refreshments.

It would be a mistake to regard the officers' mess of those days as at all comparable with the club-like centre that many of them, for entirely valid reasons, have become today. The object of the exercise was quite different. The mess was the home of the bachelor officers, and in the days when marriage allowance became admissible only at the age of thirty, and when in most line regiments there were subalterns with fourteen or more years' service waiting for a captain's vacancy, there were enough of them to create their own rather closed society. Married officers dropped in for a glass of sherry before lunch, sometimes in mid-morning, and regularly dined in on guest nights, but the mess belonged to the bachelors, ranging from misognynist majors to susceptible (but financially impotent) second-lieutenants. There were characters in the family, as in others, but if the battalion was a good one the officers' mess made a happy family contributing to the general tone, and the CO was therefore at pains to keep it together by discouraging premature matrimony and, sometimes, overseas secondment[2] as well.

Social life in the mess was a rather formalised version of social life in the equivalent strata of society outside. Officers were taken for granted as gentlemen in a meaning of that word which has now almost vanished, and they found it no great hardship to behave as such in the society of their peers. The forms that were laid down were strictly observed, but they were not unduly repressive and were for the most part accepted without much question. Those who objected soon left.

Mention of young officers is a reminder that although they held the rank of lieutenant or second-lieutenant they were never referred to as such; generically they were subalterns, socially and on duty they were *Mr* and to their correspondents *Esq.*

Dinner for mess members was a commanding officer's parade for which one turned up in time, and to spare, and properly dressed – in mess kit. It took place four or five times a week, and although one could warn out, to do so more than once or twice a week would invite a reprimand

2. This would mean a tour of duty with the Royal West African Frontier Force, the King's African Rifles or Sudan Defence Force or some other such colonial force. It would not mean posting to the foreign battalion over which the officer had no control and the CO little if any more.

from the senior dining member or senior subaltern. With so many regularly dining in mess, one was likely to find better company in it than outside, and this prevented dinner becoming a bore.

There was a guest night once a week when married officers dined in, the band was in attendance and the buglers blew the mess call. Civilian guests wore tails and decorations and the loyal toast was drunk according to regimental custom. When the General dined in he did so as a guest like anybody else and there was no departure from normal. After dinner he might demonstrate his virility by taking on a subaltern at cock-fighting but more usually he played bridge, sometimes billiards fives.

Much more formal was the GOC's luncheon in mess during his annual inspection, when the mess caterer and staff were on trial with the rest of the battalion. All officers assembled to be introduced, and they registered personality according to their natures in anticipation of the General having, in due course, to countersign their confidential reports. Amongst the items subject to inspection was the collection of training manuals which every officer had to keep amended to date. I do not remember them ever actually being inspected but I do remember feverish work with scissors and paste in order to insert a year's worth of amendments (copious amendment lists were issued almost monthly) into *King's Regulations.*

Cock fighting – alternative variation.

On supper nights (Saturdays and Sundays and sometimes, but by no means always, Wednesdays) those of us who supped wore dinner jackets and the orderly officer wore patrol uniform. Separate arrangements were sometimes made in another room for those who wished to sup early or late without changing. 'Dining dirty' was the phrase used. By no means everyone went away at the weekend (at least one officer per company was required for Sunday church parade) but most had a night out on Saturdays.

The impression created by most plays and films about army life in those days is that officers spent most of their time in mess kit and a befuddled state of alcoholism. There was *not* a lot of heavy drinking in mess. It would have been severely dealt with if there had been, and mess bills, regularly scrutinised by the PMC, would have given the game away very quickly. Moreover, most of us could not have afforded it.

Ladies only entered the mess on ladies' guest nights, which were few and far between, and there was no ladies' room. Odd though this may sound today, there was little desire for one, and the fiercest opponents of the idea would have been the single officers who would have regarded it as a threat to the privacy of their home. We enjoyed life and there were high spirits in plenty; perhaps a good deal of the enjoyment was derived from the ordered life that we lived.

The ordered life was, of course, largely controlled by the discipline of mess rules and the ordinary self-discipline of the members. It is interesting to learn that mess rules existed as early as 1811 and bore a very distinct resemblance to those of today – or perhaps that should be put the other way round! In the Royal Artillery mess at Woolwich it was ruled 'All dogs to be prohibited the mess room under penalty of a bottle of wine'. Neglects in dress and correct behaviour were similarly penalised and 'blasphemous, nasty and obscene conversation and swearing' was also forbidden. I wonder if this last rule was enforced strictly in the Peninsula, in the retreat to Corunna for example. Perhaps the issue was avoided by swearing in Spanish. It is clear that a high standard of behaviour in the mess was insisted upon in those days just as it is expected today.

It is also clear that considerable progress had been made in the previous thirty odd years for it was stated in *Advice to Officers of the British Army* published in 1782 that 'if you belong to a mess, eat with it as seldom as possible, to let folks see that you want neither money nor credit. And when you do, in order to show that you are used to good living, find fault with every dish that is set on the table, damn the wine, and throw the plates at the mess man's head'. Scarcely behaviour to endear oneself to one's brother officers!

All of us at some time or other must have longed to smack an ill-behaved child – someone else's of course. A saving grace of an officers' mess was that there simply was not room for an ill-behaved member. He was soon cut down to size or, if he still did not conform, he soon packed his bags for some other regiment or service or some colonial force where he might learn wisdom. There was no room for misfits or angry young men. As a senior officer has said, it was 'an extremely happy and pleasant communal life in circumstances of considerable comfort'. He might have added, taken all round, cheap at the price too.[3]

At first, membership of the officers' mess was voluntary. Later it became compulsory and because mess life was deemed, even in those days of purchase, to be too expensive for some officers without private incomes, the Regent's Allowance was introduced. This was not, as is sometimes believed, intended to enable the poorer officer to drink the loyal toast. It was an allowance granted to help towards the overall expenses of running a mess. General Order of 26 June 1818 issued from Horse Guards makes this quite clear and laid down that the purpose of the allowance was 'to place the mess of each regiment on the most respectable footing and to give a general extension of the advantage and benefit it is calculated to afford by enabling every individual officer to be a member of the regimental mess... The whole sum must be applied to the reduction of the daily expenses of the regimental mess for the comfort and accommodation exclusively of those officers who attend and most particularly for that of the junior part of the regiment'. Nothing could be clearer than that. Of course in those days messing allowance did not exist, or rations in kind which today help to reduce the cost of living to a good extent, and the Regent's Allowance which was often dipped into to help subsidise expensive entertainment in the dining room might in a sense be likened to that. But quite clearly it was not intended just to cover the glass – or bottle – of port that an officer might feel inclined to imbibe in drinking the loyal toast!

One expense to which officers then were not put was the purchase and maintenance of mess kit. The attire for dinner was full dress which was a normal item of an officer's outfit; as normal then as it was abnormal in the thirties, although an officer attending a levee then was expected to appear in full dress, even if he had to hire it! The Gunners appear to have started the fashion of mess kit. After the Crimea they moved into a 'stable jacket'. Readers of old copies of *Punch* will have seen the type of thing illustrated as the walking out dress of the other ranks. A bell hop's jacket, overalls and the lot topped off by a little pill-box hat, worn with PT kit by cadets at Sandhurst until the last war. The jacket was cut away like an Eton jacket but buttoned to the neck. Gradually this developed into the waistcoat and bow tie dress of today worn by the majority of regiments and corps, although the cavalry still retain the type of jacket but caught together

3. Daily messing charges, considering the standard, were not exorbitant; drinks and cigarettes were cheaper than outside; and there was block membership of the local golf club, tennis club, yacht club, etc. so that a mess member could enjoy such sports at a subscription of about 2/6d a month each.

THE OFFICERS OF THE BLUES, REGENTS PARK BARRACKS, 1882

Back row: Hon. Oliver Montagu (Major), John Brocklehurst (Captain), ?, ?,
Middle row: '?, Alfred Egerton (Major), F. Strickland-Constable (Lieutenant), ?, Sir John Willoughby, Bart. (Lieutenant), Lord Arthur Somerset, (Major), ?, Alexander Gordon (Lieutenant), ?, ?,
Front row: Hon. Lionel Byng (Lieutenant), R. Weir (Riding Master), ?, David Milne-Home (Lieutenant-Colonel), Fred Burnaby (Lieutenant-Colonel), Lord Kilmarnock (Major), Evelyn Atherley (Captain), Lord Binning (Lieutenant).

only at the neck and not buttoned all the way up. To balance the expense of mess kit, full dress has passed away and officers attend at investitures and such ceremonies in service dress or No.1 dress (more familiar to the older officers as blue patrols, the uniform worn as orderly officer to inspect tea buckets at the cookhouse at five o'clock!).

In many respects, life in an officers' mess has changed little in the past 150 years or so. New members now, as in 1811 and certainly 1931, are expected to behave 'as though they were guests in a country house showing the same courtesy to other members as they would under those circumstances'. But then how many country houses and country house parties exist nowadays – except among the pop-singing and pot-smoking community? There is one definite difference and that is the frequency of mess meetings. In the early nineteenth century in the RA mess at Woolwich they were an annual affair. Now they have to be held quarterly and often provide a source of entertainment far better than the entertainments' officer on the committee provides.

Like dinner nights, these are CO's parades, but unlike dinner nights the CO invariably presided and directed the course he wished 'his' mess to pursue. Nowadays they tend to resemble political meetings; proposals, counterproposals, compromise proposals, a lot of 'belly-aching' off the point and officers displaying their powers of oratory to impress but to no purpose. Sometimes they sadly lack direction from the chair. There was even talk of a mess where the CO sat in the body of the kirk and voted like everyone else! A little too democratic even for the 60s! In the old days the CO was very much in the chair, assisted by the PMC and mess committee. The accounts were rapidly passed. 'They've been on the notice board for days and if you haven't bothered to read them we're not going to waste time now. Proposed? – seconded? – right. Next'. Next might be a point arising out of the minutes of the last meeting and... 'No. I've said it before and I say it again we are *not* going to have dogs in my mess. Next.' Next might be proposed an entertainment – a mess ball or cocktail party. All would be enthusiastically in favour but not quite so unanimous about raising the wind. 'Can't we have a grant from mess funds?' Just like modern politicians wanting a 'grant from the government' (i.e. us taxpayers) for some hairbrained scheme. 'Of course' says the CO, 'the mess will find halt and the rest will be made up by members at the usual mess guest rates. Should not come out at more than a quid a pip. Agreed? All right? Right, next'.[4]

4. This brief and totally imaginary impression of the conduct of a mess meeting is deliberately exaggerated to illustrate that in those days as much was decided by the CO's direction as by interminable argument and a vote. It was – and still is – laid down in regulations that the CO's authority was at all times and in all places paramount. At a mess meeting it was often exercised as a sort of benevolent dictatorship but with the best interests of the mess at heart. At least that is my possibly hazy recollection of mess meetings attended as a subaltern.

OFFICERS OF THE HOUSEHOLD CAVALRY, 1961

Balcony – Left to Right:
Lt.A.H.G. Broughton (RHG), Lt.G.R.P. Coles (LG), Lt.C.P. Baring (RHG), Lt.N.W.S. Paravicini (LG), Lt.G.J. Ward (RHG), Lt.D.G. Morgan-Jones (LG), Lt.J.S. Crisp (RHG).
Steps – Left to Right:
Maj. (QM) D.F. Robarts, M.B.E. (RHG), Capt. (DoM) W. Jackson (LG), Capt. (RM) W.L. Thompson, M.B.E., D.C.M. (LG), Vety.Lt.Col. P.W. Dean (RHG), Surg.-Maj. D.H. Matthews (RHG), Lt.A.A.L. Wills (LG), Capt. T.G. Coombs (RHG), Capt. H.A.M. Pyman (LG).
Front Row – Left to Right:
Col. The Hon. J. Berry, O.B.E. (RHG), Field Marshal The Lord Harding of Petherton, G.C.B., C.B.E., D.S.O., M.C. (LG), Maj.-General Sir R. Howard Vyse, K.C.M.G., D.S.O., D.L., J.P. (RHG), Maj. I.B. Baillie (LG), Maj. H.S. Hopkinson, M.B.E. (RHG), Maj. R.I. Ferguson (LG), Lt.Col. A. Meredith Hardy, M.V.O. (LG), Capt. J.A.C.G. Eyre (RHG).

In fact it was laid down in *King's Regulations* that the CO should 'lend his countenance' to any officer who wished to opt out of an entertainment which he considered too expensive. Not many officers, however, exercised the right to borrow the CO's rather expensive port-style countenance but if they did their wish was respected – and they were probably made entertainments' officer at the next meeting just to see how entertaining they could be on nothing. And so it went on. In a very brief and usually pretty diverting time business was completed, strangely enough very satisfactorily and efficiently. Like learning, too much democracy is a dangerous thing. Then it would be ... 'Press the tit, Clanger, you're nearest and we'll have a drink while the papers are auctioned, and don't everyone bloody well bid for *Punch*'. *Punch* was the CO's choice but CO or not he had to pay heavily – more heavily for it, in fact. The young officers liked to raise the bids to the moon with no intention of getting stuck!

In those days there would be a large selection of 'shiny' papers and if an officer happened to figure in one it would be liable to cost him drinks all round. There were the *Sporting and Dramatic, The Illustrated London News, Sphere, Tatler, Bystander, The Field, Country Life,* and *Punch, Blackwoods* and maybe another magazine or two, the *Strand* for example. They were legion and the mess would take in several. The auction would go with quite a swing and would help to swell the paper fund with the aim of making it as self-supporting as possible. Woe betide any officer caught smuggling a magazine to his room, and for anyone committing the heinous crime of 'pinching' one, words could not be found. The chap was a complete cad sir, better he leave the regiment!

Mess meetings were usually held before lunch at an hour reasonably calculated to enable the ball and chained good conduct men among the marrieds to get home in good time for lunch, and for the less fettered to have one or two and be a bit late and blame it on the mess meeting; but not *too* late. 'The colonel has been home for *hours*. I saw his car'. It was all one to the subalterns. Their appetites needed no whetting and they could not afford much of the other sort of wetting.

* * * * * * * *

No book dealing with the officers' mess would be complete without a reference to the mess staff. The official staff was usually very small – the mess sergeant, mess corporal, the cook, a couple of mess waiters or wine waiters (NOT barmen please), the silverman, and that was about all. The odd GD (general duty man) was provided and from time to time the mess got its fair share (unfair share as far as the President Mess Committee, PMC, was concerned) of pressed gang labour in the form of miscreants doing 'jankers'.[5] On band nights and regimental guest nights a few of the more intelligent batmen would be pressed into service as waiters. They were selected not so much for their ability to wait as to fit the mess uniform, usually a sort of Victorian flunkey's outfit – blue tail coat, black and yellow striped waistcoat, blue trousers (not breeches!) and white gloves.

The mess sergeant was king behind the green baize door and little short of it in front. There were seldom bad mess sergeants. They would not last long if they were. They would not be tolerated as long as a poor sergeant in a rifle company. Usually they were very good, and men of vocation. No soldier with a field-marshal's baton in his pack would 'go for' officers' mess sergeant. But the man who did and whose ambition it was – he would be a volunteer, not detailed by roster as was the sergeants mess caterer – would end up very likely a lot better off than many an RSM or RQMS sweating on a QM commission.

His kingdom was the kitchen and back premises, the 'public rooms' and games room if there was one, and the subalterns would let him in. He'd train batmen to wait at table, lay the table with silver and so on but would not teach them how to 'bat'. The senior batman and the officers themselves would have to do that. He was a man who knew on which side his bread was buttered – no 'marge' for him and if, in compensation for renouncing any ambition to be RSM, he happened to receive nice Christmas presents from the local tradesmen, who would blame him as long as he did the job he had dedicated himself to do – to look after *his* officers better than those in any other mess in the army. They were indeed great men, mess sergeants, and much loved by the members of the mess. A mess sergeant of the 1st Seaforth Highlanders was so highly thought of by 'his' officers that he was accorded the exceptional honour of being dined out in their mess, in quite recent years.

The mess corporal was his Second in Command (2IC), not always so devoted to his trade. He might only be doing a stint as mess corporal whilst still having hankerings to go back to duty and aim for Regimental Sergeant Major (RSM). On the other hand he might accept officers' mess corporal as his ceiling and a far pleasanter one than corporal in a company. The mess corporal in the Ramrods was rather in that category. He had a talent for mending watches and his bunk was often surrounded by saucers full of watches in various stages of repair – rather like santals round a curry dish. Nowadays he would probably be a warrant officer in REME.

The waiters would take alternate hours or days of duty and would be ever alert to answer a ring on the ante-room bell. What *ever*? Well Captain Corcoran, hardly never – not alert. Rather like the Windmill Theatre in the war the bar in the officers' mess hardly ever shut. By the word 'bar' is meant the cubby-hole where the cigarettes and ready-for-use drinks were kept. This did have opening times and it also had closing times but these were very generous and could be extended. If there was a little party on and the bar officially closed at eleven it could be kept open, but a surcharge would be levied on drinks served after that hour. Perhaps sixpence up to midnight, a shilling from then to one o'clock and so on. The wine waiters did not mind much, in fact they almost aided and abetted because the fines went towards extra duty pay or a Christmas bonus. Guest nights, of course, were exempt!

5. Defaulters undergoing confinement to barracks (CB) were said to be on 'jankers'.

The silverman had no such perks although for him work was rather like painting the Forth Bridge. It was an endless task, probably only increased by modern amalgamations of regiments. He worked under a glaring electric light bulb in a windowless cell like the strong room of a bank which, after all, it closely resembled in nature. Not only did he clean all the silver to perfection but he also knew the history of each piece and consequently was quite an expert on regimental history as well. He would, like the company storeman, have several good conduct stripes up his arm and would be accorded similar respect by young soldiers who called him Mr.

Such then was the sort of mess and life into which His Majesty's well and truly beloved Charles Oswald Littlewart was about to plunge as a second lieutenant.

A SUBALTERN'S DREAM.

ii

2/Lt. Littlewart joins the Ramrods

What's in a name? What indeed; Major Philip Littlewart might have given the answer that it could be a lifetime of minor purgatory. He had suffered from a misguided choice by his parents. Too few devoted parents when choosing names for their offspring give proper thought to the possible consequences and handicap they may be imposing on the defenceless child.

This must surely account for the number of men one meets who are commonly called Ted or Bill or Joe when their initials do not include a letter that could possibly cover such names. Their actual names, bestowed not without good reason to the parents' way of thinking, are Aloysius Ferdinand, Cyprian or something as exotic, so Ted, Bill, and Joe are no more than nicknames.

Some parents are led astray by hopes of expectations, handsome legacies from wealthy godfathers and the wretched boy may go through life as A.S.S. Jones only to see the money go to the RSPCA to endow a home for donkeys, retired after a lifetime's service in rough seaside stations.

Others like to commemorate stirring historical events the classical example of which was J.S.F.E. Mulliner. Admirers of P.G.Wodehouse will at once recall that the initials stood for John San Francisco Earthquake (NOT fire) but at least he could be called John in his own right. Not so fortunate was the elder son of that eminently worthy and loyal couple, Albert and Victoria Raper. In a rush of patriotic fervour they christened him Verdun Dardanelles. This so preyed on him in later life that he played safe and entered a monastery.

Major Littlewart, who had been retired at a time when fortunately he was too old to be recalled in 1939 to hinder the war effort, had suffered under the first category. In expectation of riches untold from his godfathers he had been christened Philip Innes Leadbetter. Innes went bankrupt after a hoped-for coup in the city had failed, but at least he also failed by the skin of his teeth from going to prison. He didn't leave a bawbee, much to the indignation of Philip who had always wanted to know what a bawbee was. Something akin to a geegaw he imagined. However, he was never to learn from Innes. Leadbetter did better – for himself. He chased a rainbow to South Africa and made a pot of gold which, however, ignoring his duty as a godfather he kept entirely to himself. Consequently the miserable boy went through his early school life to no purpose or or ultimate benefit known inevitably but affectionately as the Pill. When he joined the army he hoped to rid himself of this millstone and decided to drop the Leadbetter. This did not really help. Philip was rather a quiet self-effacing type and, it must be admitted, tending towards being a bible-puncher and so instead of the Pill he became 'Pi' or 'Padre' Littlewart. Consequently when his and his wife's nuptial labours were blessed he was determined that his son would not suffer under a similar burden. He would NOT call him Sue![1] He called him Charles Oswald. Harmless? Surely. But

1. The American singer Johnny Cash popularised in the 1960s a song entitled 'A boy called Sue'. It tells the tale of a boy christened Sue by his father who, knowing the tough life he was going to lead, (said father apparently deserted his wife at the font) thought that giving him the additional obstacle of a ridiculous name to overcome would make him even tougher to face life's struggle. In later years they meet in a Saloon in Jacksonberg and Sue promptly does his best to kill his father who reciprocates with an equal enthusiasm and cuts off a piece of his ear. Eventually they call it a draw; Dad explains the reason for his name and they shake hands (soft music off). But Sue decides then and there that if he has a son he'll call him Bill or George, anything but Sue. The fallacy in the father's reasoning was to imagine the boy would call himself Sue and not Bill or Ginge.
Major Littlewart of course had no knowledge of the song – Johnny Cash was not born in his day, but Major-General Littlewart was only too well aware of it from constant repetition on his son's record player and at times when he looked with some disgust at the long-haired effeminate creature dressed in pastel shades of mauve he could think of no more appropriate name for him than Sue.

it never penetrated Padre Littlewart's somewhat slow moving mind that COL could be read as the abbreviation for Colonel – a rank he believed did not really exist as he himself had never seen a Colonel in the flesh. Only exist in the army list, what? Nor did CO strike him at the time as in any way peculiar since commanding officer had been an unattainable rung in his unambitious promotion ladder. Not so young Charles; he 'clicked' almost as soon as he could read, and the discovery lit the fire of his ambition – to be a CO! – a COL!! – even a Maj-Gen!!! Why not?

Incredibly he did achieve this ambition[2] but it always rather saddened him that he was never known as Ted, Bill, Joe, Charlie or, and he rather shuddered at the thought, Ossie. Owing to an unfortunate trait in his character from an early stage in his army career he became known as 'Clanger' Littlewart.

* * * * * * * *

It started at Sandhurst when being interviewed for his choice of regiments. First choice had been his father's old regiment, but the regimental representative informed him with a sigh (Clanger was never to know whether of regret or relief) that there was not a chance of a vacancy for some years. His second choice was his county regiment and they happened to be Light Infantry.[3] The interviewing officer had to inform Littlewart that his county regiment also was fully subscribed but he suggested kindly but perhaps rashly 'Why not join us – the Forty-third?'. 'Oh', replied Clanger, 'the Ox and Bucks[4] would be all right, but frankly I am not all that keen on light infantry as such.'

Well, at least he was honest, but his keenness was never put to the test because, not so strange to relate, it transpired that no vacancies existed in any other light infantry regiments. As he had not committed himself to a third choice he found himself when passing out, and surprisingly well up the list too, directed to the –th Foot, the old Rusty Ramrods. Had he not passed out so well he might have found himself directed to the Jam Stealers![5]

The –th Foot had earned their nickname in the very early days when after literally an age of peaceful and regrettably indolent service in the West Indies they were called upon to fire a volley in anger. The result had not been an unqualified success and left no alternative but to go in with the bayonet. This they did with such vigour and valour that they carried the day and were quite content – nay proud – thereafter to bear the nickname.

In fact the Ramrods were as good an unpretentious county regiment as any but were at that time, as sometimes unaccountably happened, experiencing a wave of unpopularity with the cadets at the Royal Military College (RMC). So in due course Littlewart found himself gazetted to

2. The tale is told in naval circles of the officer who kept petitioning their Lordships about his promotion – or more correctly his non-promotion. Eventually they surrendered and he was granted an interview. The great First Sea Lord himself deigned to see him and asked him what was the trouble. The officer explained, 'Well sir, as you may know, when the half-yearly promotion lists appear there is always at least one chap in the lists of whom the whole Navy says "Good God! how on earth did he come to be promoted?". Well sir, in the next list couldn't I be that chap?'.

   Many of his contemporaries unkindly claimed that Littlewart had become a major-general because the Army Selection Board had adopted a similar principle.

3. Light Infantry of course do everything 'at the double' and have a form of short cut to drill movements which are in consequence a confounded nuisance on any massed ceremonial parade. It would seem that, at any rate in the estimation of one cavalry officer, they also have a lighter approach to life in the mess than what he describes as the 'Stuffy rank-conscious heavy infantry'.

   He must have been unlucky in his experience of heavy infantry messes.

4. If you really wished to insult an officer of the Forty-third beyond all pardon the thing to do was to refer to his regiment as the 'Ox and Bucks'. It was no less than shootin' talk. The Fifty-second, who in 1881 amalgamated with Forty-third to form the Oxfordshire and Buckinghamshire Light Infantry, also felt pretty strongly about the Forty-third. Indeed there is a story of an officer of the Fifty-second who always shaved with a cut-throat which he stropped 52 times every day and when he came to 43 he spat.

5. The affectionate name for the Royal Army Service Corps, now living as part of the Royal Corps of Transport, as far as its transportation role is concerned, and in the Royal Army Ordnance Corps as regards its other duties.

that regiment,[6] the senior of two entrants. He was later to bless his Sandhurst clanger for he got his Captaincy several years before he would in most other regiments.

After spending his first month as a 2/Lt. on leave getting his kit together the day arrived when Littlewart duly reported for duty to the home battalion. He was surprised not to see his fellow 2/Lt. on the train – the only possible one in the day to the somewhat remote part of England where the battalion was stationed. The reason became clear on arrival when he discovered that, having misread and later mislaid his joining instructions, he had arrived a day early – on the day of the administrative inspection in fact. However, he was dealt with not too unkindly and not many days later was invited to be the guest of the officers at dinner – in short to be 'dined-in'.

HIS FIRST MESS KIT
'Gad Sir, it was just made for you!'

6. Few people will have seen a 'Commission', the document, that as far as officers were concerned called one to the colours. So for interest and historical record Littlewart's commission is reproduced below, losing a lot, of course by not being in script and lacking the royal seal signature and so on but it reads as follows:

'GEORGE THE FIFTH by the Grace of God of Great Britain, Ireland and the British Dominions beyond the Seas, King, Defender of the Faith, of India, etc. To Our Trusty and well beloved *Charles Oswald Littlewart* Greetings. We, reposing special Trust and Confidence in your Loyalty, Courage, and good Conduct, do by these Presents Constitute and Appoint you to be an Officer in Our Land Forces from the twenty-ninth day of January 1925. You are therefore carefully and diligently to discharge your Duty as such in the Rank of Second-Lieutenant or in such other rank as We may from time to time hereafter be pleased to promote or appoint you to, of which a notification will be made in the London Gazette, or in such other manner as may for the time being be prescribed by Us in Council, and you are in such manner and on such occasions as may be prescribed by Us to exercise and well discipline in Arms both the inferior Officers and Men serving under you and use your best endeavours to keep them in good Order and Discipline. And We do hereby Command them to Obey you as their superior Officer, and you to observe and follow such Orders and Directions as from time to time you shall receive from Us or your superior Officer, according to the Rules and Discipline of War, in pursuance of the Trust hereby reposed in you.

Given at Our Court, at Saint James's, the twenty-sixth day of January 1925 in the fifteenth Year of Our Reign. BY HIS MAJESTY'S COMMAND'

It is signed at the bottom right by the CIGS and the Secretary of State for War and at the top left by His Majesty personally.

iii
Doing Poojah

The Ramrods had no particular initiation ceremony for newly-joined 2/Lieutenants. Nothing, that is, that might be considered more suitable for the induction of an entrant into some primitive secret society in darkest Africa. Their aim was merely to give the new arrival a thoroughly good evening, which would be enjoyed by all, and at the same time to take their measure of him; to note his table manners and his ability to hold his liquor; to discover his interests and conversational fluency; to see how he reacted to a bit of friendly ragging; in fact to assess his rating as an 'officer and gentleman' and his suitability to join the family.

Littlewart, though not overburdened with brain, did not lack the social graces and survived the test well. The consensus was that 'he'd do'. His companion was also judged to be all right though a bit damp behind the ears, but he made a greater hit with the men for no better reason than that he had a double-barrelled name. The other ranks loved a double-barrelled name! 'Mr.Littlewart 'e's all right', they said but (with bated breath) 'Mr.Looootine-bell ...'. You could see the hyphen. In fact Mr.Lutine-Bell brought little to the battalion but the glamour of his name as the men were not slow to find out. The members of the mess were no slower and were soon to claim that he was so wet that you could shoot snipe off him!

In all regiments it was customary to call on the mess on first joining[1] and to sign the visitors' book noting in the remarks column 'on joining'. Similarly, on leaving, the book was again signed and 'p.p.c.' entered. Two cards were dropped on the mess, one for the CO and the other for the officers. In the 1st bn the East Surrey Regiment as a formal gesture of joining the officer 'took salt' with the regiment with due ceremony and signed the 31st Huntingdonshire Salt Book. It is so titled because the battalion trace their ancestry back to the 31st Huntingdonshire Regiment in 1782.

Receiving accounts by letter from his friends who had joined other regiments, Littlewart discovered that their experiences were similar to his but there were some exceptions and variations on the theme of being welcomed into the Regiment.

'Clanger old boy', wrote Barrow who had joined the Dorsetshire Regiment, 'you've simply no idea. They made me do Poojah.[2] After dinner – and I must say they did me proud, couldn't have done more to the CIGS, champers bubbling out of my ears – but after dinner ... there was Poojah. I had to crawl across the floor of the ante-room escorted by a major whilst all the subalterns kicked my bottom and walloped me with billiard cues, until I arrived in front of our silver sphinx where I had to do obeisance. The sphinx commemorates the capture of Fort Marabout in the campaign against the French in Egypt. It is part of our hat badge. Yours, too, for that matter; all regiments who took part in that campaign have it but only *we* have Marabout inscribed on it. Yah! you old Ramrod! All the time during the route march the other officers thumped the Poojah accompaniment on the Moplah Drum, captured in the Malabar campaign of 1921-22. Quite

1. In a similar manner the officer called on the mess when being posted and joining the other battalion or the regimental depot.

2. Doing Poojah (Sanskrit 'puja') is descriptive of certain rites in the Hindu religion. It is not unnatural that the regiment whose motto is *Primus in Indis* should describe its ceremony for the initiation of young officers as Doing Poojah.

modern music, eh? I can tell you it was some act. When they reckoned I had done enough Poojah I was allowed up and made to drink a concoction in the Dorset colours. Red, yellow and green. I don't know if this was to ease the pain or put me out of my misery once and for all – it nearly did, a close run thing – but I survived! But my bottom's still sore. However, I am jolly glad I joined; they are a topping lot. How did you get on? Drop anything?'

In the 2nd Bn The Seaforth Highlanders the newly joined 2/Lt. was spared Poojah but later in the evening was made to drink a concoction made of all the drinks in the mess, known as a Ross-shire Buff (in 1881 when county titles were given to regiments of foot the 72nd and 78th Foot amalgamated to form the Seaforth Highlanders with the secondary title Ross-shire Buffs, The Duke of Albany's). It is as well that this *coup de grace* was not administered till late in the evening for the wretched officer could not have been fit for more than to be piped off to bed.

Cartwright had joined the Royal Norfolk Regiment and wrote to Littlewart to say 'Honestly, Littlewart, I don't know whether they were pulling my leg or what but they told me all about the Chilianwala Table. This is our colossal dining table. Apparently after the battle of Chilianwala the bodies of twelve officers of the regiment and the RSM of the 24th Foot (don't know how a bloody Welshman muscled in on the act) were laid out on it. They said you could still see the blood-stains – beer stains more like – but I could not see any though I pretended to as that seemed the best thing to do. But it is one hell of a table. It has twenty leaves, each on four folding legs and is 39ft.7½in. long and on guest nights they try to jump over a leaf but no-one's ever done it yet. It's simply not on.[3] I suppose you managed to acquit yourself with your customary glory at your night. How did it go?'

Unforgivable mistakes made by a 2nd Lieutenant on his first Guest Night in the Mess. See page 117.

As we know, Littlewart's behaviour at his dining-in left nothing to be desired but he was glad that he had never had enough brain to think of going to 'The Shop' and becoming a Sapper. If he had, the ceremony of 'The Gauge' would have been his lot. One who did undergo it describes it as follows.

It must be remembered that the gentlemen cadets leaving the Shop, instead of being dispersed in twos and threes around the army as from Sandhurst, were split into two parties and sent off to

3. That famous Victorian, Fred Burnaby of the Blues, could have done it. He could vault a billiards table and at school could jump 4ft.6in. with another boy on his back.

A Subaltern's Quarter in an Officers' Mess. *circa* 1910.

See page 113.

try their luck with the Gunners and Sappers respectively. Later this became three when the Royal Corps of Signals was formed. From the Sapper point of view this meant that twice a year a batch of hopeful young officers, about twenty strong, would arrive at Chatham from the Shop, and as the YOs course lasted three years there were some hundred officers living in the HQ Mess.

About three weeks after their arrival each batch was dined-in at a special guest night. Each YO was taken into dinner by a senior officer and given the full treatment. By the end of dinner he would be in fine form, even though slightly tinged with apprehension at the 'gauging' ceremony which was known to lie ahead. The Gauge in ordinary life is an imposing piece of furniture standing behind the top table in the mess dining room. Four massive legs, which would not disgrace a billiards table, support a solid oak top, and there are two intermediate shelves. It is used by the mess steward to dispense the wines, port and liqueurs during dinner. On a batch night, however, it is spirited away to reappear after dinner in the centre of the large ante-room which has been specially cleared for the occasion.

As the newly commissioned young officers emerge from dinner they are herded into the billiards room and divested of jacket, waistcoat and spurs. When the senior batch are ready for the fray the first victim is called forward. On entering the ante-room he is ordered to stand on the 'bum-warmer' and announce his full name. When this has been satisfactorily given the head of the batch gives the command 'Gauge Mr.Snodgrass'. Snodgrass is then seized by the gauging party and forced into a horizontal position, hands stretched forward. In this posture he is fed into one end of the Gauge, between the middle and lower shelves and his hands are grasped by the party stationed at the other end of the Gauge. He is drawn forward so that his head and shoulders are exposed and then beaten with rolled up newspapers, as a blacksmith might attempt to shape a piece of iron. He is then partially withdrawn and the same treatment applied to the other end. The head of the batch then asks 'Does he fit?'. If Snodgrass has caused any bother to the gauging party the answer will probably be 'No', in which case the procedure is repeated until the cry goes up 'He fits!'. Snodgrass is then hauled smartly out of the Gauge and projected across about six feet of the ante-room into a settee, from which he is rescued by other subalterns and given a drink. He is now 'in'.

This process continues until all the new batch have been gauged. On this occasion it pays to be one of the junior members, since when your turn comes the gauging party may have run out of papers. This is fine unless they turn to *Country Life* instead.

The only recorded occasion on which the even tenor of this ceremony has been disturbed was in the nineteen-thirties. A batch arrived from the Shop full of rugger players and self-confidence. Four of the senior batches were away at Cambridge and the new boys reckoned they were more than a match for the other YOs. They hatched a plot to rush the ante-room when their first man was called in, seize the head of the batch, and gauge *him*. This plan was put into operation and the opening rush caused some consternation. However, it is agreeable to record the very quick reaction of the other officers who were sitting round in the ante-room ready to watch the fun. As one man they rose to their feet in support of the senior batch, and fell on the rebellious youngsters.

It was a very chastened and well gauged batch who paraded under the senior drill instructor at 0630 the next morning for the traditional hour's bayonet fighting before breakfast.

Newly-joined officers are dined into the Royal Regiment at a special dinner at Woolwich at the end of their initial course at the Royal School of Artillery. The young officers and their fathers are guests of the regiment, and the hosts include the Master Gunner, the Director, Royal Artillery, one or more colonels-commandant, and as many of the commanding officers of Gunner regiments to which the young men have been posted. (This really is being welcomed into the family.) In Littlewart's day, when the officers then joined their units they might face the ordeal of drinking the health of Cardinal Puff. Although not called upon to do this in the Ramrods, Littlewart became a master of the art! In fact it was his party piece.

To return to the less technical infantry and a musical ceremony, in the King's Liverpool Regiment the officer being dined-in is allowed to conduct the band. Put baldly this conveys little of what might happen. As friend Needham might say, 'The finibilities are imposite'. Perhaps the conductor may choose his own music. *Colonel Bogey?* dedicated to the CO? Unwise!

iv

Dining customs and ceremonies

When Littlewart joined, the custom in the mess was to have dinner every night except Wednesday, Saturday and Sunday. On these days there was supper. This was only a slight concession; it differed from dinner in that it was a running meal within tight limits instead of a formal one, and dinner jackets were worn instead of mess kit. Littlewart wondered in later life how on earth he had afforded the laundry bills. All those stiff shirts, to say nothing of wear and tear on mess kit which on hilarious guest nights was not treated with the respect due to such an expensive outfit.

Of course, it was always possible to order early or late supper at a small charge for extra messing and in some messes it was possible to 'dine dirty behind the screen'. For this a screen was placed behind the vice-president's chair in the dining room and officers who wished to go out in the evening could partake of the same meal as served to those dining in mess kit, and were able to sneak off when they liked. Silence had to be observed; one should be neither seen nor heard but this rule was not always observed and a smart reprimand from the president relayed by the mess sergeant would follow.

Once a week the loyal toast would be drunk, and married officers might take the opportunity to dine in and be credited with an attendance against the quota expected of them in the month. Once a month a full regimental guest night would be held and the dining table would be graced by many handsome variations of mess kit.

In fact, mess kit did not vary very much between infantry regiments the main difference being in the colour of the facings and waistcoat or cummerbund in tropical mess kit and the number of buttons on the jacket cuff. They were not so resplendent as cavalry mess kits which in turn could seldom hold a candle to Indian Cavalry! The Forty Third, now the 1st Royal Green Jackets, are distinguished by the unique touch of wearing a white waistcoat and a white bow tie. The Essex Regt., now a component of the 3rd Royal Anglian, share the fashion of wearing a white waistcoat but stuck to the black bow tie but in tropical mess kit the 2nd bn were distinguished by sporting a purple cummerbund – to the occasional indignation of the Chaplains who might claim that they were trespassing on holy ground. The origin of Pompadour purple as the colour of the battalion's facings is supposed to stem from the eighteenth century. The Colonel of the battalion at that time petitioned the King for permission to adopt blue as the regimental colour for his facings. The King would not approve, so the affronted Colonel adopted the favourite colour of Mme de Pompadour whose favours he had once enjoyed – or at least claimed to have. In the circumstances, therefore, it is the clergy who are poaching on the layman's preserves rather than the reverse. Pompadour purple remains the regimental colour of the 3rd Royal Anglian today.

The 2nd Royal Green Jackets are distinguished or indistinguishable by virtue of wearing no badges of rank in mess kit, a custom shared by Guards officers who, however, do have a system of distinguishing chevrons worn on the sleeve above the cuff.

The 3rd King's Own Hussars had a nice way of 'decorating' and identifying the junior subaltern. At dinner he would wear a silver collar presented to the regiment by Lady Southampton in 1772. This collar is still worn, but by the kettle drummer on parade.

It did not take long to get into mess kit.[1] It took longer in India with many feet of cummerbund to wind round the waist, even with the assistance of a bearer, and by a quarter to eight or so most officers would be assembled in the ante-room. In due course, after an aperitif or two, it would be time for dinner and the dinner call would be sounded. This would usually be the familiar 'Officers' wives have puddings and pies, and sergeants' wives have skilly', but several regiments have traditional calls usually tracing their origin to that great breeding ground of regimental customs, the Peninsular war.

The Sherwood Foresters, for example, sound a call played on cornets arranged from the tune of a carillon of bells of a nunnery in Spain. The Mother Superior made a present of the tune to the Colonel of the 45th in grateful appreciation of his men having spared her charges from whatever normally happened in those days.

The Officers' Mess Call –
Officers' wives
Get puddens and pies,
A Sergeant's wife gets skilly,
But a private's wife gets nothing at all,
To fill her empty belly.

(Skilly - a thin gruel)

1. Unless one was unlucky enough to have experienced the difficulties of a certain 2/Lieutenant of the Devonshire Regiment. In order to don mess wellingtons and overalls the boots had first to be strapped into the overalls which were then rolled down concertina wise until the calf of the boot was accessible. It was then easy to step into the boots and pull up the overalls. The unfortunate 2/Lt. Hanks had allowed himself only the minimum of time to change into mess kit and was in a bit of a hurry. He plunged his feet into his wellingtons to find that his new batman, who until then had not put a foot wrong, had contrived to put the boots into the wrong trouser legs, so instead of being just in time the wretched Hanks was a little late and incurred his CO's displeasure.

The Senior Subaltern welcomes the new addition to the Officers' Mess. circa 1890.
'Knew yer father – hope you're as good an officer as he was'.

See page 113.

These Officers are drinking their pre-dinner sherry in the Mess ante-room at a School of Instruction. circa 1895.
Mess dress, left to right: 'Heavy Cavalry', Royal Field Artillery, Foot Guards, Rifles, a Scots Regiment, Royal Engineers.

See page 115.

The call to summon officers to dinner in the 1st and 3rd Royal Green Jackets, the second mess horn (so named from the early days when bugles were made from cows' horns) is the old Light Division Assembly call.[2] Its purpose then was to get the men on parade and on the march. Now it summons the officers to sit and dine together!

In the Royal Highland Fusiliers the duty bugler plays the dinner call and the duty piper *The Heroes of Vittoria.*

To commemorate their service as Marines, the 50th Foot (7th Marines), later to become the Queen's Own Royal West Kent Regiment, were always 'piped' to dinners and the custom was observed at regimental guest nights. A bugler, fifer, and a side-drummer would parade and after the officers' mess call the fifer and drummer would 'pipe' dinners.

The mess sergeant having got the office from the senior officer present, or whoever is the officer responsible to give the word, then announces that dinner is ready by proclaiming loudly and clearly 'Gentlemen dinner is served'. There is an orderly procession to the dining room led by the CO, guests and their hosts and senior officers. The rest then follow, not necessarily in order of seniority although certain regiments do prefer it that way. During the move to the dining room the band plays *The Roast Beef of Old England* and continues playing until all are at their appointed places, either seated or ready to sit down as soon as grace has been said.[3]

The order of march is sometimes varied. In the Scots Guards the picquet officer leads the way and also acts as president for the night. The procession of officers of The Buffs from the ante-room into the dining room was invariably led by the senior officer dining and, as invariably, the most junior subaltern brought up the rear. Woe betide the latest arrival from Sandhurst who, even all unwittingly, failed to go in as 'tail end Charlie'.

In the Royal Berkshire Regiment, now amalgamated with the Wiltshire regiment to form the Duke of Edinburgh's Royal Regiment, the officers were preceded by two drummers playing a roll. In the dining room they separated and each drummer stationed himself in the centre of opposite sides of the table and continued to play until all the officers were seated. They then left the room. One of the drums was captured in the Crimea from the Russians; the other was taken prisoner from the Germans in the 1914-18 war when the officers determined that before it was over a companion for the Russian must be acquired. It was, and so there is a drum for each side of the table.

If you happened to be dining with the Royal Leicestershire Regiment you would find the silver and crockery enclosed within a strip of black crepe, in the shape of a lozenge in the 1st bn and a rectangle in the 2nd bn. This custom known as 'laying the Crepe' keeps alive the memory of General Wolfe under whose command the regiment had served in Canada and who was held in great esteem by them. Strangely enough, Wolfe did not serve in the Royal Leicestershire Regiment but his father did.

The dinner table is ruled by the president and vice-president seated at the ends of the table. Normally they are appointed by roster from the living-in captains and subalterns working towards the middle from the top for president and the bottom for vice-president. There are exceptions – in the Border Regiment the orderly officer was the president and the next for duty was vice which must have been very consoling for the memory. For an extra special regimental guest night a more

2. It would appear that this call owed its origin to General Crauford's Standing Orders for the Light Division, Section 1 'Preparations for the March', an extract from which reads as follows, 'The third horn, or assembly, will sound a quarter of an hour after the second, upon which the Captains will march their Companies to the parades ...'. Thus the third horn has become the second mess bugle call. Its significance remains the same; to get the company on the move – to the dining room.

3. The saying of grace is not a universal custom. It depends largely on whether a padre is dining in, and on the CO's inclination. Grace in the 1st bn East Surrey Regiment was said and the custom followed was as follows. The president called for silence for grace before and after dinner by striking a double knock with the silver hammer presented by four officers of the Indian Army. As soon as everyone was seated grace was said by the president who also said grace at the end of the meal. The usual grace was 'For what we are about to receive (have received) thank God'. If a chaplain were present he was asked by the CO to say grace before the meal and all officers remained standing.

senior officer may be detailed to act as president for the success of the dinner rests largely in his hands and although the mess sergeant directs the detailed activities, if the CO considers that the dinner did not go off as it should, it is odds-on that the president will be invited to the orderly room for a chat with the CO next day!

The conduct of the meal as far as behaviour is concerned is in the hands of the president and his vice. Officers may not address mess servants direct; an indent for their requirement must be passed through the usual channels – that is to say to the president or vice! The silver must not be handled before the port has been circulated; a lady's name must not be mentioned; shop[4] and religion are forbidden topics of conversation. Failure to observe these customs may involve the offender in the summary award of drinks all round imposed by the president. On the other hand, there was no rule, written or unwritten, which prevented embarrassment to many who were forced to listen to eulogies on huntin', polo, pigstickin', fishin' and shootin', in which sports the listener could not possibly afford to take part on £15 a month. Nor did conduct become too riotous even after the port had been circulated, and feats of 'skill' such as trying to balance as many wine glasses as possible one on top of the other were not exactly encouraged, although if stocks were low they might not be too strenuously discouraged since breakages had to be paid for by the offender at several times the value of the item! (Six times the value in Royal Scots Fusiliers.)

Water Colour by Cecil Cutler – Green Howards, 2nd Battalion, Bordon 1889.

The chief duty of the president and vice is the ceremony of proposing the loyal toast, dealt with later as a special subject. Suffice it to say that it must be performed without a hitch. It is not a good thing for everyone to stand with glasses raised at the ready waiting in a deathly hush for the band to strike up. The toast having been successfully drunk it is the time for coffee and cigars or cigarettes and liqueurs. Port and cigarettes do not blend too well and it was sometimes the custom that if cigarettes were allowed at all they had to be Turkish! In the past at this stage the bandmaster would enter the room and sit beside the president or CO and be invited to take a glass of port and cup of coffee, whilst the band played light music under the direction of the band sergeant. Two glasses of port was the ration and then the bandmaster had to return to exercise his authority and wield his wand again.

Normally the coffee and cigars would be ushered in by the mess sergeant in response to a tinkle on the president's bell or some such signal, but a pleasant variation existed in the 12th Royal

4. It is interesting to note that it is laid down in the Royal Artillery Mess rules of 1811 'No regimental business is to be discussed at the mess under the penalty of three bottles of wine for each separate offence'. It rather sounds as if some Gunner officers of those days were ultra-keen on discussing regimental business.

Lancers. When the port had been round twice the junior subaltern present got up and went to the colonel and asked 'Colonel, may we please have coffee and cigars?' Receiving the colonel's assent he relayed it to the orderly officer who was seated at the end of the table with a silver bell, presented before the 1914-18 war by Major Eustace Loder.[5] This he rang and the mess staff duly brought in coffee, cigars, liqueurs, cigarettes (and probably snuff).

In the Royal Highland Fusiliers a cigarette box and lighter and a snuff box are placed in front of the CO and 2IC for the sole use of themselves and their guests. This is in continuation of a custom followed by 2nd Battalion The Highland Light Infantry in which this privilege was accorded the CO. Why is not recorded. Perhaps he had his own favourite brand of cigarettes and snuff!

In the Seaforth Highlanders the rule that there should be no smoking until the loyal toast had been drunk was overlooked when the then Prince of Wales (later to become the Duke of Windsor) dined in the mess. He liked to smoke between courses and the mess sergeant and waiters had to be on the *qui vive* ready to offer him a cigarette or provide a light.

In due course of time the CO will give the lead to the ante-room and officers and their guests will follow suit at their leisure. Then the vigil of the president and vice begins, for they must be the last to leave the table. This may be dreary for the vice if the president is busy exchanging reminiscences of long service in bad stations in the East with an old friend! However, if he is so engaged he will probably have compassion on his vice and give him permission to leave.

What might be called the official ending of the meal would coincide with the end of the band programme. The various regimental marches of the guests and of the hosts and in some regiments the national anthem would be played, the band would fall out and the officers fall in for 'any other business'. Many regiments have their own particular way of bringing the programme of music to an end. In the 12th Royal Lancers, after the various regimental marches of the guests have been played, the band plays the Regimental Music. This consists of the three regimental hymns, the slow march *Coburg* and the quick march *God bless the Prince of Wales.* The three regimental hymns are a Spanish chant, the Sicilian Vespers, and the old Russian Imperial Anthem. The origin of the custom is lost in the mists of antiquity but tradition has it that they were presented to the regiment by Pope Pius VI when the regiment was in Italy in 1793. Their regimental music is played while the officers are still seated at table and its conclusion with the anthem was a tip to the colonel that a move to the ante-room would not come amiss, particularly for those whose powers of continence may have been impaired by service in India.

In the 2nd Battalion The Border Regiment it was the custom that immediately after the officers had left the dining room the band would play the *Chinese Airs* and *Rule Britannia* before going on with the rest of the programme. The latter commemorates the service of the 34th Foot with the fleet in the eighteenth century whilst the former recalls service of the 55th Foot in China in 1840/2.

But possibly the pleasantest custom was that of the Northamptonshire Regiment. After playing the regimental marches the band played *Rule Britannia* and *God bless the Prince of Wales* and finally as a unique feature the bandsmen sang in unison, 'Bon Soir'. This custom is still maintained by 2nd Royal Anglian Regiment. It is a trifle anomalous because with *Bon Soir* the night really begins!

5. It was quite customary in the old days for officers to present silver to the mess on promotion, leaving the battalion to join the other battalion, on winning a prize in the Irish Sweep or for any reason he saw as a fit excuse. This custom is seldom observed now. It is too expensive a gesture. An interesting example of the custom is the Silver Fox which is placed on the dining table at guest nights in front of the junior subaltern in the Queen's Lancashire Regiment. It was presented to the regiment then the Loyal Regiment (North Lancashire) by a famous CO, Lt-Colonel Walter Hill. He was a madly keen foxhunter and on leaving the regiment presented this lovely piece of silver. It is placed in front of the junior subaltern simply because the famous Maida Tortoise is always placed in front of the CO. (For more of the Maida Tortoise snuff box see Chapter VI.) There is no connection with the Aesop fable — a fanatical fox hunting man would not acknowledge the existence of Harriers!

   Colonel Hill presented a silver fox to the regiment for no other reason than that he loved fox hunting and he loved the regiment. What better justification is needed?

V

Toasts

'I looks towards you.'
'I has your eye.'
'I likewise bows.'
'I winks and blinks according.'

So might Mr. Jorrocks and his friend Jemmy Green have proposed a mutual and reciprocal in health before setting about ruining it in copious draughts of port or claret. In the modern methods of proposing health or toasts there is nothing so colourful or evocative. Toasts in Gaelic are often attractively worded but they need to be translated before the Sassenach can appreciate them. A toast in Welsh is included later in this chapter but there is nothing attractive in the translation which is both rather anatomically crude and, in the circumstances in which it was proposed, undoubtedly rude.

The custom of drinking healths – many centuries after to be called toasts – was observed in the ancient world although it took the form rather more of pouring a libation and circulating a loving cup. It was natural that the Romans should introduce it into England and once established the custom was enthusiastically indulged in by other subsequent uninvited guests to our shores, sometimes to their considerable advantage.

Vortigern, who appeared to have learnt nothing from the Roman presence and whose director of intelligence was scarcely up to the job, lost a large portion of his kingdom to Hengist owing to being poorly briefed on the customs of the latest tourists to Britain. He fell into the classic trap of wine and woman. Too much wine in the form of toasts and too much woman in the shape of Hengist's daughter. He was literally caught on toast. When the evening – a royal guest night – was well advanced, Rowena entered the hall bearing a loving cup and pledged Vortigern with 'Lauerd King wacht heil!', (or words to that effect) to which he learnt it was proper for him to reciprocate with 'Drinc heil'. He picked up the drill readily and after a few more *wacht heils* and *drinc heils* Rowena had little trouble in hooking and landing him and leading him to the altar. If a list of presents had been drawn up there might have appeared the unusual entry 'from the bridegroom to the bride's father – all Kent'.

The origin of the word toast to mean a health is somewhat confused. One explanation lies in the custom of putting a piece of toast in the loving cup which was being circulated. When the cup had finished its round and arrived back at the originator he was expected not only to finish the wine – no great hardship – but also to eat the toast. Many a man not unnaturally preferred not to and so 'dunking' went out and in the strange way that often happens the word toast became transposed and applied to the drink.

Another explanation is given in an account of an incident which happened in Bath in Charles II's day.

A celebrated beauty of the time was in the Cross-Bath and one of the crowd of her admirers took a glass of the water in which the fair one stood and drank a health to the company. 'There was in the place a gay fellow, half fuddled, who offered to jump in and swore tho' he like not the liquor[1] he'd have the toast.' He was restrained, but the process of drinking a health to your lady-love became known as drinking a toast. Drinking the health of the King became such a nuisance in the Restoration that the King was forced to express his displeasure at 'such vicious debauched and profane persons ... giving no other affection to us but in drinking our health'.[2] This may be an explanation why many regiments do not drink the loyal toast at all. The usual explanation given for this exemption is that the loyalty of the regiment has never been in doubt, for example during the Jacobite troubles. On the other hand, the loyalty of certain regiments may have been open to question, for they were ordered to drink the Majesty's health every night; others once or twice a week. Nowadays whatever the practice it is merely a matter of tradition.

Other more interesting reasons for abstention exist, as for example in the King's Shropshire Light Infantry. George IV, when he was seeking a divorce from the Queen, became very unpopular. On one occasion when he was attending a theatre in Brighton, a town which was very pro-Queen, rioters became violent and there were fears for the safety of his person. However, officers of the 85th Foot (2nd Bn The KSLI) who were there protected him and he escaped unhurt.

Soon afterwards the King dined with the officers and to show his appreciation of their loyalty gave them dispensation from drinking his health or standing for the National Anthem.

In fact, in the very early days of the British Army the Sovereign of the day made quite a practice of giving regiments this dispensation from drinking his health. For this reason the loyal toast was not drunk by the 1st The Royal Dragoons, the 11th Hussars and the Royal Fusiliers to whose officers George IV said 'Gentlemen pray be seated, there is no need to drink the health. The loyalty of the officers of the Royal Fusiliers could never be in doubt'.

There must often have been cause for doubt, however, for a custom still observed today in the Royal Regiment of Artillery owes its origin to such suspicions. Nowadays 'proper precautions are taken to ensure that the loyal toast is drunk by all officers'. These precautions include the following.

When the port is going round, officers who do not take wine are invited to fill their glasses with water; when standing for the toast, officers hold their glass with forearm extended in front of them, for all to see; and in the headquarters mess and other messes where dinner is held by candlelight, the main lights are switched on for the period of the loyal toast so that the president may see that all officers drink the toast.

Other regiments have earned the exemption by action on the field of battle; the 5th Royal Inniskilling Dragoon Guards for their loyalty at the battle of the Boyne; the 9th Queen's Royal Lancers for their part in the attack on Buenos Aires in 1806 and the Durham Light Infantry who on the conclusion of the campaign against the Caribs in the West Indies were also granted the designation 'Faithful', in 1772.

The 1st Cameronians (Scottish Rifles) adopted a compromise based on religious scruples. The president remaining seated said 'Mr.Vice the King', to which Mr.Vice replied 'God bless him'. The wine was then circulated but no toast was drunk. This custom is associated with the origin of the regiment in 1689. The Covenanting days of the 17th century can never be forgotten in Scotland. In those distant times there was a strict sect of Presbyterians led by Richard Cameron. After the accession of William and Mary this sect agreed to raise a regiment from amongst its numbers under certain conditions, such that 'there should be no profane talk, no drunkenness and no drinking of toasts'. The regiment took its title from its founder; whether it still fulfills the other conditions or not it certainly does the last – no toasts have been drunk ceremoniously.

1. The usual drinks for the loyal toast are port and madeira and brown sherry. In India, in deference to the climate, whisky and soda is added to the list, usually served in individual carafes, ready mixed and of about half-pint capacity. This the officer decanted into his wine glass.

   An honest tee-totaller is of course allowed water, and his opposite, the budding Master of Wine, may with the permission of the president, recharge his glass with the wine he has been drinking.

2. Queen Elizabeth I also expressed her displeasure when encountering one of her courtiers, who was clearly in an advanced state of euphoria. To her demand to know what he had been doing to explain his condition he said 'I' faith Madam drinking your Majesty's health'. 'So I thought and I am sorry for it' replied the Queen, 'For I never fare worse than when my health is drunk'.

Some regiments are not total abstainers but only drink the loyal toast on special occasions. The 2nd Bn The Duke of Cornwall's Light Infantry drink the toast only once a year on the Royal birthday. It is believed that this custom dates from the siege of Lucknow when wine was not available to drink the Queen's health and so it was only drunk on her birthdays. From this circumstance the custom was born and lived on.

The Royal Welch Fusiliers is another regiment which drinks the toast only once a year, on St.David's Day. This custom originated in 1797 when the mutiny at the Nore was raging. The naval mutineers invited the Fusiliers to join them. They refused flatly and in fact submitted through their sergeant major an address to the Colonel expressing their loyalty and devotion to the King. This was conveyed to his Majesty, who is reputed to have granted the officers the dispensation from drinking his health at the weekly guest night, stipulating that it need only be drunk once a year. Naturally St.David's Day is the chosen time. The other ranks, who had so remarkably expressed their loyalty, received no reward such as free beer from the Royal cellars on that annual celebration!

The Royal Welch Fusiliers. Ceremony of Eating the Leek. *See page 116.*

In the Life Guards the toast was and is only drunk on very rare occasions, rather when the spirit, and perhaps the wine, moves the officers, and then it is proposed by the senior officer dining. The Highland Light Infantry, who had been exempted for their undoubted loyalty drank the toast only when a distinguished guest was present. The presence of royalty was the only excuse for drinking the toast in the 17/21st Lancers and the King's Royal Rifle Corps.

A number of regiments which habitually drink the toast do so seated. This almost certainly indicates that the regiment was among the number who saw service in the eighteenth century on board warships and thus started life virtually as marines. In due course of time they rejoined the 'Pongo'[3] fold but retained the naval custom of drinking the toast. Many are the ingenious explanations brought forward to explain the naval custom but little support seems to be given to the obvious one – there simply was not enough head-room to stand smartly erect; a Uriah Heep type crouch would be about the best that could have been achieved.

But this explanation – marine ancestry – can scarcely be applied to cavalry. In those days as in these there were no horse marines to whom to tell the tall tale! Nevertheless when, on very special occasions only, the toast is drunk by the 4/7th Dragoon Guards it is drunk seated. It was also drunk every night by the pride of Scotland, the Royal Scots Greys, a privilege reputedly granted them by the monarch.

3. The origin of this sailors nickname for soldiers appears to be unknown, possibly some type of Gibraltar ape?

St. David's Day Ceremony

After a picture by S. T. Dodd in possession of the Regiment.
The Regimental ceremonial of eating the leek takes place on St. David's Day, which is celebrated by the Royal Welch Fusiliers, wherever they may be.

See page 116.

A very different reason is given in the 1st Bn Royal Ulster Rifles. The health was drunk every night, a custom apparently originating from the grant of the Regent's Allowance, and in order to avoid undue inconvenience the habit developed and continued for the officers to remain comfortably seated whilst they enjoyed the indulgence of the Regent!

An unusual but very logical procedure exists in some regiments of proposing the toast seated but drinking it standing. In the Royal Lincolnshire Regiment (now the 2nd Royal Anglians) the procedure was and is as follows. As soon as the decanters have been placed on the table and the band has stopped playing, the president rings his bell for silence, the vice president also rings his bell. The president, without rising but raising a decanter in his hand, says clearly and loudly enough for all to hear 'Gentlemen, charge your glasses; we rise to drink the health of His/Her Majesty'. The president and vice then start the decanters from their respective ends of the table and normal conversation is resumed. When the president is satisfied that the wine has been circulated he again rings his bell. The vice president also rings his bell and rising to his feet and holding his glass breast-high says 'Gentlemen the King/Queen'. All then rise and stand during the playing of the first bars of the National Anthem. The officers then repeat the words 'The King or Queen', and drink the toast.

Very similar procedures existed in the Loyal Regiment, the Cheshire Regiment and the King's Own Royal Regiment, the differences lying almost solely in the wording of the toast. The Green Howards have a slight variation in that the president hammers three times with his gavel, rises and bows towards the vice president and says 'Mr.Vice the Queen'. The vice president rises and bows towards the president. Both then sit and circulate the port. When all the glasses are full the vice president hammers three times with the gavel and says 'Gentlemen the Queen'. After which the procedure follows the normal.

Yet a further variation on this theme existed in the Worcestershire Regiment. Before passing the wine the president said 'Mr.Vice we will drink the King'. When everyone had filled their glasses he rose, all other officers rising too, and said 'Gentlemen the King'. The officers repeated 'The King' and drank the toast to the usual musical accompaniment. It will be observed that the president 'hogged' the stage and that Mr.Vice had no particular role in the ceremony.

To round off this discourse on the drinking of toasts seated or part seated and part standing, mention must be made of the customs observed by the Royal Sussex Regiment. In the 1st bn (late 35th Foot) the toast is drunk seated in commemoration of the days when the regiment served as marines. It is drunk *every* night by the president and vice and anyone else who might wish. The president and vice help themselves to wine and, seated, make the normal toast and the band, if present, plays. Both then instruct the waiter to 'Pass the wine please'. The wine is passed and each officer individually and in his own time toasts 'The Queen'. On guest nights all officers drink the toast. It is believed that when the Regent was living in the mess the wish was expressed to him that the loyal toast might be drunk every night. He, realizing the expense this might cause officers, apparently suggested that a system be devised whereby officers could do so if they wished. The procedure noted above was introduced.

A similar system was adopted by the East Surrey Regiment and the Leicestershire Regiment for ordinary nights but the wine was passed by the officers.

In the 2nd Bn The Royal Sussex Regiment (107th Foot) the toast is drunk standing since, as their number denotes, they were raised some time after soldiers served as marines. But in both battalions the wine is passed by mess servants, which is surely a unique practice.

In 1953 Queen Juliana of the Netherlands was appointed Colonel-in-Chief of their regiment. Later she visited the regiment at Tidworth and dined in the mess. The loyal toast was drunk seated but Queen Juliana's health was drunk standing. After this she turned to the Commanding Officer Lt-Colonel J.B.Ashworth and said 'If you drink your own Queen's health seated then you must drink mine also seated'. So was another custom given birth.

In the majority of regiments the toast is proposed and drunk standing, the 'normal' procedure being on the following lines. The president starts the wine circulating, the vice follows suit and all officers fill their glasses. The president then calls for silence by tapping the table with his knuckles, the stopper of a decanter or some other such means[4] or by ringing his bell. He then rises and proposes 'Mr.Vice the King'. Mr.Vice stands in his turn and replies 'Gentlemen the King'.

'The King, Mr. President' – This scene of Peninsular War days is reproduced from a painting by Matania. It depicts the occasion in the Regiment's history when only two officers survived the battle and for the first time the King's Health was given in the unique manner now in current use in the Gloucester Regiment.

It once happened in the Buffs that a youthful Mr.Vice, who had dined rather unwisely, did not give the correct reply to the president but slouched slightly sideways in his chair slurred 'Show him in, show him in'. It was he, who was shown out smartly to await the CO's pleasure at orderly room next day. Assuming the correct reply is given, all rise with their glasses held at breast height whilst the band breaks into the first six bars of the National Anthem. That concluded, the officers pledge 'The King' drink the toast and resume their seats. This or something very like it was the normal drill. Scottish regiments, of course, added a little Gaelic flavour of their own. But there are many variations of even this simple procedure hallowed by tradition.

The first to be mentioned must be that of the 1st Bn The Gloucestershire Regiment. During the Peninsular War in a particular battle the officers had suffered such casualties that there were only two survivors left to drink the King's health at dinner. The president proposed the toast in the normal manner 'Mr.Vice the King' but Mr.Vice, as there were no other members present, had perforce to reply 'Mr.President the King'. The loyal toast has been drunk in this manner ever since and the occasion has been immortalised in Francis Matania's splendid canvas which hangs in the battalion mess.

As already mentioned, in the Worcestershire Regiment Mr.Vice took no part in proposing the toast. Nor did he in the East Yorkshire Regiment; the president both proposed and replied. The origin of this custom is believed to be the regrettable inability of a Mr.Vice to rise to the occasion and consequently the president had to carry out both parts of the proposal himself!

4. Gavels are a popular instrument. In the officers' messes of both battalions of the Queen's Royal Regiment a gavel made from the timbers of HMS Excellent was used. This ship was the descendant of HMS Charlotte in which a detachment of the Queen's fought as marines at the 'Glorious First of June' which is an unusual battle honour for soldiers.

The President of the Mess committee of a Royal Regiment inspects the seating plan and table before a regimental Guest Night.

See page 113.

No such risk was run by the 12th Royal Lancers. The toast was proposed by the CO himself or the senior member dining, who simply rose and proposed 'Gentlemen the King'. That the CO should become *hors de combat* would be unthinkable!

In both the Border Regiment and the Essex the vice president omitted the word Gentlemen. Littlewart was stupid enough to enquire the reason from the Essex vice one night when *he* was a bit *hors de combat* having been dined well. He received the rather scathing reply that 'in *my* regiment we assume we are all gentlemen'.

Normally the toast is drunk from ordinary port glasses and the visitor to a strange mess would be wise to drink only a little, for another toast might be following on the heels of the loyal toast with no opportunity to recharge glasses. But in the 2nd Bn The Dorsetshire Regiment it was the custom to drink a bumper toast with no heel taps from 1802 to their disbandment in 1949. This custom originated in Gibraltar when the first Duke of Kent, father of Queen Victoria, was Governor and Commander-in-Chief. On Christmas Eve the troops of the Garrison mutinied and marched on the Governor's residence with the intention of killing him. The 54th stood firm and dispersed the mutineers; a second attempt was made on the Duke's life two days later and was similarly dealt with by the 54th. In gratitude the Duke presented the regiment with a magnificent silver punch bowl which is still the centre piece of the silver on the table at dinner nights of the Devonshire and Dorset Regiment. It is in commemoration of this presentation and the preservation of the Duke's life that the Bumper Toast was drunk. Before the wine is passed the president, seated, says 'Mr.Vice a bumper toast'. The Vice responds 'Gentlemen a bumper toast'. The wine is then circulated and the rest of the procedure is normal. The toast was also drunk with no heel taps by the King's Own Royal Regiment and the Royal Berkshire Regiment.

* * * * * * * * *

These are but some of the variations attending this apparently simple ceremony. Further confusion can be caused for the unwary officer in a strange mess from the fact that in those days when regiments were two battalions strong, one method of drinking the toast was adopted in the 1st bn and another in the 2nd bn, as has already been noted in the case of the Royal Sussex Regiment.[5] Whilst amalgamations have lost much of the old regimental glamour they have also ironed out a few of these problems! Furthermore, in many regiments the procedure might differ between ordinary nights (i.e. when the toast was drunk every night) the weekly 'band' night and the periodical regimental guest night. On top of that there would be second toasts immediately after the loyal toast to Colonels-in-Chief and so on, as well as special regimental toasts commemorating a battle honour or some big event in the regimental history. The relief of Ladysmith, we are asked to believe, by the Gloucestershire Regiment is one. According to the Regimental Customs Book (see Chapter IX) if a guest night falls on 28th February the officer commanding the admin company is compelled to dine off bully beef and biscuits and drink water until the loyal toast has been drunk. Then he will rise and give the toast 'Gentlemen Ladysmith is relieved'. To judge by the company commander's title this must be one of the newly instituted customs possibly originating in a desire to cause discomfiture to an officer who may have been a bit of a gourmet if not gourmand. The west country seems to abound in strange customs but one, the validity of which there is no disputing, is the toast drunk to the French Army by the Devonshire Regiment. After the battle of Bois de Buttes on 27th May 1918 the 2nd Bn The Devonshire Regiment (together with some other British battalions) was awarded the Croix de Guerre. To commemorate this award a health was drunk each year on that date. After drinking the King's health the RSM marched in carrying the old Battalion Camp Colour, in the centre of which was hung the original Croix de Guerre medal, and stood behind the mess president while the band played the Marseillaise and the officers stood. The health of the French army was then drunk.

5. This difference was often not restricted to the procedure for drinking the loyal toast. Sometimes the two battalions of a regiment could be as different as two regiments – possibly not even on speaking terms. The officers' uniforms might be quite different. In fact if an officer was posted from the 1st Bn The Essex Regiment to the 2nd Bn (The Pompadours) his service dress would require considerable alteration. Colonel Walter Hill of the Loyals, who has already been mentioned, was so keen on hunting that amongst other things, recalls Brigadier G.A.Rimbault CBE DSO MC 'he re-dressed our 2nd bn in the 1920s giving the officers a cavalry look. For our plain khaki breeches he substituted fawn whipcord which could also be worn with a black hunting coat. These were made by Dege and were magnificent but very expensive. He also replaced our stiff cheese-cutter hats with the softer 10th Hussar type and our swagger canes with short sticks, which were known as 'pipp' sticks. (Please I play polo)'.

   Presumably all this had to be approved by the dress committee but it is doubtful whether a CO could get away with it today!

The Seaforth Highlanders preferred to drink a toast to themselves. After the playing of the piobaireachd (pibroch) the pipe-major would follow the mess sergeant back into the dining room and receive a quaich of whisky from the senior dining member whereupon he would give the regimental toast 'Deoch Slainte na chabar faidh agus ciad mhic coinnich' or in English 'Drink the health of Cabar Faidh and the first regiment of MacKenzie'. After giving the regimental toast the pipe-major would drain the contents of the quaich then turn it over and kiss the base. In response all officers present would raise their glasses and say 'Slainte'.

In the Camerons a similar ceremony took place but the pipe-major gave no specific regimental toast and in the Scots Guards the privilege of draining the quaich was not reserved for the pipe-major alone. After the port and coffee have been passed round a piper plays and when he has finished he is given a quaich full of whisky and proposes the Gaelic toast 'Deoch Slainte na ban Righ'. Many people south of the border are acquainted with these Gaelic toasts and though not understanding them know them to be imbued with kindly spirit and well wishing. This can scarcely be said of a certain toast given in Welsh.

After the last war the Durham Light Infantry were stationed at Piddlehinton and at a big guest night the Royal Welsh Fusiliers were their guests. After the usual toasts, Major Lloyd Davies got up, as representative of the guests and toasted their hosts. A fine upstanding figure, he raised his glass and as the guests were all fluent Welsh speakers gave the following toast in Welsh 'Twll dyn pob sais'. The RWF officers rose to the occasion, raised their glasses, repeated the toast and sat down again. The DLI felt bound to reply with a few well-chosen and complimentary words, toasted the Welshmen and sat down. The DLI colonel, curious to learn the meaning of the charming Welsh toast, asked his RWF opposite number at table to translate. He did so saying that it meant 'With respect sir, rissoles[6] to all Englishmen!'.

Not the politest way of giving thanks for the hospitality extended to them. No doubt the Geordies returned the compliment in kind and as near the bone by debagging the lot after dinner.

* * * * * * * * *

The 51st Foot (1st Bn King's Own Yorkshire Light Infantry) kept alive acts by one of their officers in the Peninsular War which in a later age would almost certainly have earned the award of the Victoria Cross. The officer was Ensign Joseph Dyas and he was considered to be 'one of the bravest soldiers in the Peninsular', an opinion endorsed by the Iron Duke himself.

On the 6th of June Ensign Dyas volunteered to lead the Forlorn Hope in the assault on the breach in the walls of the fortress of Badajos. They came under such heavy fire that the order was given for all to retire and during the withdrawal Dyas had his cap blown off and his sword shot out of his hand. Two days later he was in command of a detachment – a standing patrol – to watch for any movement between Badajos and an outpost. He was to be recalled before daylight but no order was received and at dawn their situation under close range of the French was critical. So he selected a trusty Irishman to go back to the CO for orders. The soldier correctly repeated his orders adding 'recall or no recall I'll be back in five minutes dead or alive'. Fortunately he came back alive with orders for recall. On the next day, Ensign Dyas still not having had enough, insisted, against the wish of General Houston that he should stand down, once more leading the stormers. The assault again failed. Dyas, though wounded, continued to try to scale the breach but he was up against overwhelming odds and increasing casualties compelled him to withdraw. The Duke, when he heard of Dyas' bravery, sent for him, congratulated him and offered him immediate promotion into any regiment in the Peninsular; but Ensign Dyas preferred to stay with his own.[7]

For a long time he was toasted throughout the Peninsular Army – an extraordinary recognition in itself – but in time he and his stormers came to be remembered only by their regiment. The toast of 'Ensign Dyas and his stormers' is proposed by the senior officer present shortly after the loyal toast and is drunk in silence.

6. It does not require a Welsh scholar to guess that this was not the exact word used!

7. A brave and loyal decision in those days of purchase when a famous Gunner officer did 21 years as a subaltern and is reported to have remarked that 'When a man has done 21 years as a subaltern he can never get on'. But he disproved his statement. In the next 19 he made Field-Marshal. He was Sir Richard Dacres.

The loving cup of the 14th/20th Hussars – 'The Emperor'.

01.35 hrs. A Regimental Guest Night for the Brigadier and his staff, the Master of Foxhounds and a rich landowner. (between the two great wars).

See page 114.

Another toast, owing its origin to gallantry in the field, was drunk by the 50th Foot (Queen's Own West Kent Regiment). This was to the Corunna Majors. At the Battle of Corunna the 50th Foot retook the vital village of Elvina during the 16th of January. Sir John Moore watching the counter-attack cried out 'Well done my majors'. The majors referred to were both of the 50th Foot – Major Napier, later General Sir Charles Napier, and the Hon. Major Stanhope, who was killed later in the battle. The memory of these two officers and the Battle of Corunna was perpetuated by the officers of the regiment on the guest night on or nearest to January 16th. After the loyal toast the president would rise and say 'Gentlemen, the Corunna Majors'. The officers would then rise and repeat 'The Corunna Majors' and the toast would be drunk.

The foregoing toasts were drunk by the company in the normal manner. Many regimental toasts were drunk by circulating a loving cup.

On the anniversary of the Battle of Talavera it was the custom of officers of the 48th Foot (The Northamptonshire Regiment) to hold a dinner for all the officers of the regiment in the neighbourhood. At this dinner the Talavera Cup, which was purchased by the officers of the regiment in 1877, is filled and passed round and a solemn toast is drunk to the regiment and the memory of those who have given up their lives in its service. The custom has grown that the junior subaltern of the regiment present at the dinner has to finish the contents of the cup whilst he proposes the toast. The cup holds six bottles of champagne and the subaltern on this occasion is afterwards allowed to leave the dinner table without asking the permission of the president.

This custom was kept alive on the amalgamation of 1st and 2nd bns of the regiment in 1948 and is now still observed by the 2nd Bn Royal Anglian Regiment.

Another loving cup known as the Dettingen Cup was circulated annually on 27th June, the anniversary of the battle of Dettingen in 1743, by the East Surrey Regiment and a special toast to the 1st Bn The East Surrey Regiment was drunk by each officer in turn. It is not specified whether the lucky – or unlucky depending on the point of view and Bacchanalian tendencies of the officer – junior subaltern had to drain the cup. But it was the rule in the Royal Irish Fusiliers that, when on a guest night the 'Barrosa Cup' was circulated, the senior guest had to drink the remains, often, it is said, with disastrous results. Come come, senior guest of those days, surely you could take it? even if you did require to leave the table you would never have had to ask the president's permission!

But the finest loving cup of all – and its normal place was not on the dining table but under the bed – must surely be 'The Emperor', which is circulated in the 14/20th Hussars. After the battle of Vittoria in 1813 the 14th Light Dragoons, who were leading the pursuit, captured the travelling coach of Joseph Bonaparte, Napoleon's brother, who was also King of Spain. In the coach was found a handsome silver chamber pot with the arms of Imperial France engraved on it, which may well have been a gift from his more famous brother. (Chapter XII has further information on chamber pots and their use).

The 14th Dragoons took possession of the chamber pot and it has remained in the officers' mess to this day. On guest nights after the loyal toast and a toast to the regiment's Colonel-in-Chief, HRH the Princess Anne, the mess sergeant brings in the chamber pot filled with champagne. The commanding officer rises first and drinks to 'The Emperor' and the Emperor is then passed round as a loving cup and everyone stands up in turn and drinks the same toast. Those drinking it for the first time, including guests, receive thunderous applause. When all have drunk, the commanding officer directs the mess sergeant to invite one or more of the officers dining to finish off the Emperor, and the last one, who might be the last joined subaltern or an officer about to retire, to hold it upside-down over his head on completion.

In a mess meeting book there is a minute passed at Aldershot in 1929; 'Proposed that, as before 1914, the junior subaltern be not required to finish the Emperor'. This decision was taken because on one occasion the Emperor was so full that the junior subaltern of the time passed out after finishing it and was a considerable time in being brought round. Hence the present custom of more than one officer being invited to finish it off, depending on the depth of its contents after everyone has drunk. Hence also the regiment's erstwhile nickname – 'The Emperor's Chambermaids'.

Many regiments have as their Colonel-in-Chief a member of the Royal family, and their health is drunk after the loyal toast in a similar manner, but a few regiments have foreign Royalty and the manner of drinking their health is often unusual. One or two examples are worth recording.

Mention has already been made of Queen Juliana of the Netherlands, Colonel-in-Chief of the Royal Sussex Regiment; the Green Howards also had a Queen as their Colonel-in-Chief – Queen Alexandra. After her death in 1925 and until 1942, during which period King Haakon of Norway was Colonel-in-Chief, the toast drunk in silence, was 'To the memory of our late Colonel-in-Chief Queen Alexandra'. Now their Colonel-in-Chief is King Olav V of Norway and the toast is 'Our Colonel-in-Chief King Olav V of Norway'.

Littlewart once dined with the Buffs, whose Colonel-in-Chief was the King of Denmark. The background music to the toast is, naturally, the Danish National Anthem. This is a fine, impressive piece but on the lengthy side and punctuated by certain pauses, the length of which led the uninitiated to believe the music had ended and the toast was about to be drunk. Clanger of course was misled and was raising his glass to his lips with a firm 'King Chr...' when he realised that the music was continuing and that he was guilty of the crime of applauding during movements in a symphony! Poor Clanger, he always meant well but life was always playing tricks like that on him.

This chapter can best be closed with a unique toast recalled by a past CO of the Devonshire Regiment.[8]

'It was I think' (his memory is not perfect) 'in November 1934 that several of my friends and I – then serving with the 2nd Bn The Devonshire Regiment at Connaught Barracks in Dover – went down to see the arrival of HRH Princess Marina, who was coming over to England for her marriage to Prince George, Duke of Kent. Although we were very much an unofficial reception committee our uniform must have been a sufficient passport to secure an excellent view of the Princess as she stepped ashore from the destroyer in which she had travelled. There is no doubt we were all stunned by the beauty and attractiveness of this girl from Greece.

That night at the end of mess the port circulated as usual and we all without exception filled our glasses and drank a bumper toast to 'our Princess'. The toast pledged, we threw our glasses over our shoulders with results which a subaltern could ill afford.[9]

I had the great pleasure of telling this story to Her Royal Highness when as Colonel-in-Chief of the Regiment she honoured us with her presence at luncheon in 1964. She had not previously heard of it and appeared to be delighted.'

I trust that this most popular and well loved late member of our Royal Family was not displeased with the circumstances of this truly delightful and unique toast being brought to the notice of more than that select band of officers of the Devonshire Regiment who had the pleasure of welcoming her to England with such spontaneous gallantry, so completely in harmony with the original conception of a toast.

8. Lt-Colonel G.N.B. Spencer.

9. In the hey-day of toast drinking this was the supreme gesture. The glass was broken so that no other toast could ever be drunk out of it.

TOASTING PROCEDURES

A. The loyal toast was/is not drunk as a matter of routine in the following regiments.

The Life Guards
The Queen's Bays
3rd Carabiniers
5th Royal Inniskilling Dragoon Guards
1st Royal Dragoons
3rd King's Own Hussars
7th Queen's Own Hussars
9th Queen's Royal Lancers
11th Hussars
13/18th Hussars
15/19th Hussars
16/5th Lancers
17/21st Lancers
Grenadier Guards
Coldstream Guards
Welsh Guards
The Royal Fusiliers
The 43rd Light Infantry
The Duke of Cornwall's Light Infantry
The King's Shropshire Light Infantry
The King's Royal Rifle Corps
The Durham Light Infantry
The Highland Light Infantry
The Rifle Brigade

B. The loyal toast was drunk seated or proposed seated and drunk standing.

The Royal Scots Greys
4/7th Dragoon Guards
The King's Own Royal Regiment
The Royal Norfolk Regiment (up to 1893)
The West Yorkshire Regiment
The Green Howards
The Cheshire Regiment
The Worcestershire Regiment
The East Surrey Regiment (1st bn only)
The Border Regiment
The Royal Sussex Regiment (1st bn only)
The Royal Ulster Rifles

C. The loyal toast was drunk with no heel taps in the following regiments.

The Dorsetshire Regiment (2nd bn only)
The King's Own Royal Regiment
The Royal Berkshire Regiment

D. The loyal toast was drunk in the 'normal' manner with only minor variations (if any) by all other regiments and corps.

VI

Port and a sninch

Have a glass of port Brigadier
I really think you ought Brigadier.
My subalterns will pay
They get ten bob a day.
Do have another glass of port Brigadier.

This somewhat libellous rhyme appeared in a book, which had quite a vogue in the thirties, called *Ask the brave soldier.* It is a jolly little rhyme but quite inaccurate. An officer – least of all the CO – would never have behaved in such a manner even if it had been possible, which in fact it would not.[1] The brigadier would either have been the CO's personal guest or a mess guest. In the latter case charges were always levied on a pro rata basis, often at a rate calculated on the number of pips an officer wore. The CO would have been charged with five shares to his subalterns' one or two for every glass with which he plied the brigadier. The point is made; this CO was entertaining, or trying to entertain, a personal guest at the expense of his subalterns. Such types might have been met during and just after both wars but a regular CO of such a type in the thirties would have been as rare as the dodo.

The loyal toast is usually drunk in port though it is not, as Hermione Gingold once claimed in a *Sweet and Low* revue, the only wine. You may have some Madeira m'dear or sherry, for Madeira and dark sherry kept port company on its journey round the table. In India, as already mentioned, whisky and soda was permitted, but not out of a tumbler. Port, however, has come to be accepted as the thing for the loyal toast in the officers' mess just as in the sergeants' mess port and lemon was the correct tipple for the ladies. The quality of the port might vary. The VSOP might only come out on regimental guest nights unless especially requested. For 'ordinary' nights in consideration for the officers' pockets something less expensive might be provided: the 'Boys Port' or the 'Hunting Port', still excellent but costing only a few pence a glass.

A few years after the war Littlewart was reminiscing at a guest night in his own mess with an old Indian Army friend and extolling, over the port, the good old days until his friend commented 'Clanger, you must live in the present you know. You are too much of a *laudator temporis acti*'. 'All right, perhaps I am, but *eheu fuglasses posthume posthume labuntur annas*; which my old Nabob and Wykehamist scholar I will translate for you in my inimitable FitzGerald style as "God! the price of a round nowadays". Give the port a fair wind old chap, you're holding it up.'

It was as well Clanger was not comparing prices with those ruling in his father's mess when port could be bought at 24s.0d to 144s.0d for – wait for it – a dozen. But Clanger's friend had left the Indian army on partition with a pension, had joined the Ordnance and become a colonel almost at once and so was well breeched both with £sd and rupees, annas and pice!

1. However, if the regiment had, as is most likely, an alert and observant mess sergeant, he would have seen to it that the entry in the wine sheet was in the right place. He might even have taken a tactful opportunity to whisper, in the presence of another officer, 'Beggin' your pardon sir but I presume the Brigadier is your guest'. The subalterns would *not* pay!

Whilst the port is circulating, behaviour at the table becomes freer and easier. You may handle the silver (but not try to pocket an attractive snuff box) and conversation becomes looser as overalls become tighter and bladders are tested to their elastic limit. 'Why doesn't the CO GO?' Of course it was possible to ask the president for permission to leave the room but some officers considered this too prep-schoolish and rather than lose face would sit and squirm, thankful at least that the band was not going to play selections from Show Boat. *Ol' man river* would have been the last straw!

An officer in the Scots Guards was saved this discomfort. After the port had circulated twice he could leave the table without asking the president's permission. This sensible custom probably also existed in some other regiments.

Once the port has started to circulate there is no limit to consumption beyond the officers' ability to carry it and pay – cash and carry in fact. It is reported by the Gloucestershire Regiment that during the Peninsular War 'the wine of the country being excellent we all enjoyed ourselves very much; so much so that some of us bivouacked under the table for the night'. An admirable idea, preferable to travelling home after a guest night in BAOR hopefully but suicidally up the wrong side of the autobahn.

Mr.Jorrocks, of course, preferred to be tied in his chair. Possibly the unhappy vice president of the East Yorkshire Regiment had tied himself in forgetting he had yet to give the reply to the toast! A maxim of Ensign Odoherty may not come amiss here – 'If you intend to drink much *after* dinner, never drink much *at* dinner and particularly avoid mixing wines. Port, three glasses at dinner; claret, three bottles after; behold the fair proportion and the most excellent wines'. Nowadays it is not likely that there will be a claret session after dinner but it is likely that Sir Morgan's ration of port will be exceeded.

Port, as probably everyone knows, is circulated to the left[2] but precisely why has not been satisfactorily established. 'With the sun' it is said. But the sun does not circulate although it was once supposed to, and in the heroic age when libations were poured to gods and superstitions abounded and it might consequently and reasonably be supposed birth was given to this custom, the ancients passed the wine from left to right. This gave partial excuse for the custom in the Ramrods of passing the port in the direction of Oporto. The origin of this custom is supposed to lie in an incident which occurred when the battalion was serving in the Mediterranean. The president of the night, an irascible type, absentmindedly started off the port to the right. Being taxed with his error he refused to admit it claiming that port *should* be circulated towards Oporto, just as a Mohammedan lays his prayer mat towards Mecca, and furthermore that when *he* was serving in the Med that was the way Odysseus would have passed the wine and what was good enough for Odysseus and the ancients was good enough for him and the Ramrods. And it has been ever since. It is the duty of the battalion intelligence officer to work out bearings from the president's seat at table to Oporto and decide which way the port should circulate. Littlewart got his drinks and geography mixed and drew a bead on Seville but he didn't own up and no one ever queried his decision. As a result the port in the Ramrods is as often circulated to the right as to the left.

However, it has now circulated quite enough and the Brigadier has had his fill so let us join Dr.Spooner in a sninch of puff.

* * * * * * * * * *

The habit of taking snuff originated in 1683 when the fleet brought a large quantity to England. By 1711 special classes were being held in 'The exercise of the snuff box – to be taught with the best plain and perfumed for offer to strangers, a friend or mistress[3] according to the degree of familiarity – with an explanation of the careless, the scornful and the surly pinch and the gestures proper to each'.

What a splendid sight a class of instruction, taken by a Master of Snuff, must have been. On

2. If an officer had let the decanter pass him by and then either decided after all that he would like another glass of port or realised that another toast was coming up and his glass was empty he might pass his glass to his left-hand neighbour to have it charged and returned direct. This was known as having a 'backhander'.

3. Presumably not in bed; but the thought does conjure up visions in the imagination.

regimental guest nights there might be the tentative, apprehensive, junior and field officers' pinches – but whatever the type of pinch it had to be taken with proper ceremony. The following is the drill reported by the Royal Norfolk Regiment. On regimental guest nights the snuff box is passed round the dining table with due ceremony. The president places the box between the first and second fingers of the left hand, gives a couple of taps on the lid with the knuckles of his right hand (to level the snuff in the box) then lifts the lid and takes a pinch of snuff between the thumb and fingers of the right hand, and having raised his left thumb places the snuff in the hollow formed at the base of the thumb. The lid is then closed, the box again given a couple of taps then passed to the left, the next person gripping it between the first and second fingers of the left hand and carrying out the same drill. The taken snuff is 'sniffed' and the results awaited.

One cannot resist the temptation of imagining a class of instruction of young officers being conducted in the billiards room by the senior subaltern, using the classic words of the contemporary volumes of *Small Arms Training* ... 'Watch me and I will give you a complete demonstration'.

A snuff box dated 1805 which was the property of the 2nd Bn The Royal Norfolk Regiment is now with the 1st Bn Royal Anglian Regiment. In the 2nd Bn Royal Anglian Regiment the snuff box is tapped *three* times before it is passed on whether or not a pinch has been taken.

The Gloucestershire Regiment, home of many strange and amusing customs, carry out the ceremony as follows. All officers take a thumb nail, sniff, sneeze and call together 'God bless the Dukes Sussex, Cambridge and Duchess of Kent, Patrons of this excellent snuff'. This is to commemorate the time when the regiment ran out of snuff and the mayor of Cheltenham, a purveyor, stepped in and provided the excellent snuff mentioned. This custom would seem to call for great control of the olfactory nerves and may even require to be taken with a pinch of salt. Many are the snuff boxes in existence and often still in use, of interest both artistically and historically. One has already been mentioned, the erstwhile property of the Royal Norfolk Regiment. Another infantry regiment, the Loyal North Lancashire, possessed two, both with interesting histories. The 1st bn (the 47th Foot) had one made from the silver gorgets which officers used to wear hung round the neck to denote their rank. When badges of rank were transferred to the epaulette the officers of the 47th had their gorgets melted down and fashioned into a snuff box of four compartments, each one bearing the name of one of the four most popular brands of snuff at the time. But the snuff box of the 2nd bn (the 81st Foot) must be unique. At the battle of Maida the 81st were commanded by Lt-Colonel John Kempt who was later one of Wellington's Brigade commanders at Waterloo. After the battle 'he supped off a tortoise which came into his tent.[4] He preserved the shell as a memento and later had it mounted in silver as a snuff box which he presented to the officers' mess. It is still in use. On guest nights the tortoise is filled with snuff and passed round the table after the port by the Queens Lancashire Regiment of which the old 47th and 81st are ancestors.

The 1st Bn The Queen's Own Royal Kent Regiment had a snuff box with an interesting and unique origin. On every guest night, when the table was laid, a silver snuff box was placed in front of the chair where the commanding officer was to sit and after the port it was passed round so that those would like to could take it. This is nothing unusual but the history behind it is. A certain officer had a serious difference of opinion with his commanding officer who called upon him to resign his commission – which he did. He subsequently presented the snuff box to the regiment with the special request that it be placed in front of the commanding officer at dinner nights to remind him of the 'snuff' that he had forced Mansergh to take during his service. The inscription of the snuff box reads: 'Southcote Mansergh Esq. to his brother officers of the Fiftieth Regiment as a small mark of the affection he bears for them 1834'. The snuff box is 4in. long, 2¾in. wide and 1½in. high.

The Royal Artillery mess at Woolwich is rich in silver presented by regiments whose officers have been honorary members of the mess during the time their regiments were stationed there. The first so presented is the Mortar snuff box given by the 17th Lancers in 1851. Until that date only Marines had been part of the RA Garrison. The 17th Lancers were quartered in West Square and later in the Marine Barracks. They spent only a few months there during the Great Exhibition but as a remembrance of those months presented this snuff box which is a well finished model of a 10-inch mortar with a lid over the muzzle and a wooden box in the bore for the snuff. The 9th

4. The QM department must have been slipping. Why no jugged hare?

Lancers presented a box on the lid of which is depicted a horse artillery gun in action with an escort of 9th Lancers. The work is in relief and so accurate that it will stand close examination with a magnifying glass. The only presentation from an infantry regiment, however, appears to be the Gold Snuff box presented by the Tower Hamlets Militia (5th Bn Rifle Brigade) who had a long association with the Royal Regiment and were embodied at Woolwich for both the Crimean War and the Boer War.

The Green Howards possess a gold snuff box of rare interest. It belonged to Marshal Ney and was taken from his coach after Waterloo by the ancestor of an officer of the regiment who later presented it to the regiment. On the lid is a fine miniature of Napoleon who, it is said, had one made for each of his Marshals.

There was certainly plenty of loot to be 'liberated' in those days – not by any means to be sneezed at!

'Dam' me sir, port is the only wine!'
Painting by Rex Whistler – Welsh Guards

vii
A burra tamasha in India

Do picea hubble-bubble, do picea ghi,
Sahib ke waste bibi lao, hamara waste bhi;
Toro chini toro char, Bombay bibi bahut achcha,
Sahib ke waste bibi lao, hamara waste bhi.

The above little quatrain, sung to the tune of *There was a lodging house/boarding school over the way* is probably familiar to all old Qu'Hais. It has little to do with 'The Littlewart Story'. If he had had the time he would have had more sense than to linger in Grant Road with Bombay bibis, but it must serve to indicate that we are about to go on a three weeks' trooping voyage to India; more if you happened to catch a boat that was going up the Persian Gulf to Basra delivering RAF details.

When the economic crisis of the early thirties struck, Littlewart decided that he would join the Ramrods foreign battalion, at that time in India, both to improve his financial position and widen his experience. They were expecting to finish their tour in India on the Frontier. He had little difficulty in exchanging with another subaltern who was nearing his captaincy and had his eye on the adjutancy of the home battalion which was shortly to become vacant. An adjutant in those days received extra duty pay and so the post, although it entailed more work, was sought after. Besides, an adjutant was usually considered a blue-eyed boy destined for as sure an ascent to the higher realms as Elisha.

Of the voyage Littlewart recalled little, there was little worth recalling! At Port Said he saw the sights and widened his experience (though scarcely of soldiering) no more than, if as much as, his son in the years to come was able to do in London! But at least at Port Said they were not so hypocritical as to call it art. The inevitable and unnecessary purchases at Simon Artz including one of the unnecessary purchases which no-one could resist the 'Bombay Bowler' (to be returned to the Canal on the homeward journey), did nothing to improve his financial position; nor did the depredations of the Galle Face Hotel and the jewel sharks at Colombo. He had drawn a month's advance of pay in England but that had to cover the expense of the voyage, tips to stewards and so on as well as to last him out his first month in India for he would get no more pay until the end of the month. Fortunately gin on board was about a penny ha'penny a tot so he did not reach Bombay entirely broke, only nearly so.

On arrival at Bombay, Littlewart was amazed to see a bearer, in fact several of them, waiting for him on Ballard Pier all claiming to be sent especially for him. However he eventually found the official Ramrods bearer in the throng – or perhaps more accurately his bearer by some strange instinct found him.

'Lieutenant Littlewart sahib?' He stretched the last syllable to infinity.

'Littlewaaaaart sahib'.

'Yes.'

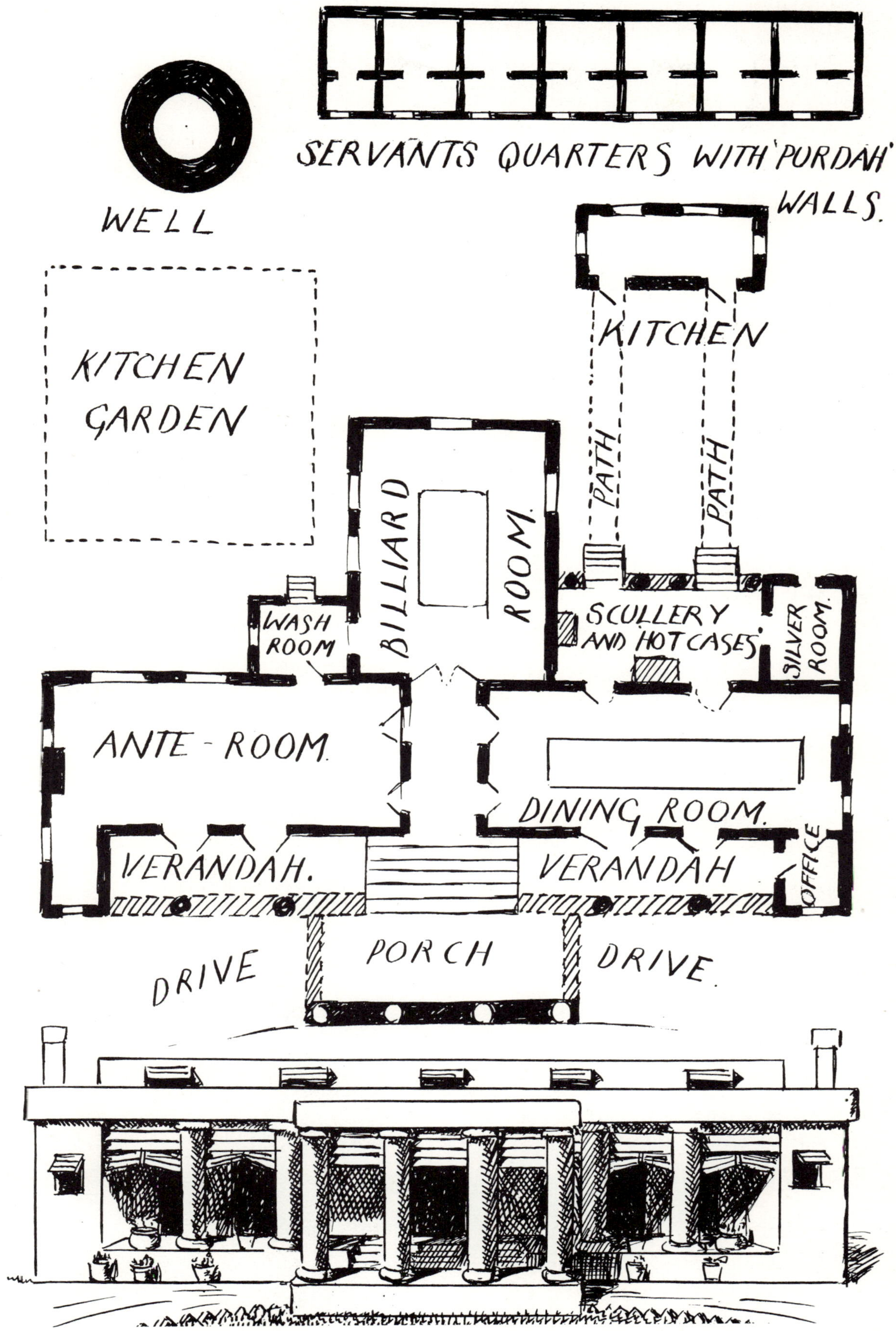

See page 116.

An Officers' Mess in India where the 18" thick walls were made of bricks and dried mud.

An interruption to the Ritual of the Tying of the Cummerbund during the hot weather in N. India.
Officers in India wore white mess jackets and cummerbunds in the hot weather. In the Indian Cavalry the cummerbund was usually the full dress lungi (turban) of a Sowar (trooper) of his regiment. The Officer revolved gracefully like a ballet dancer while his bearer secured it behind his master.

See page 115.

Pay day for the Mess servants takes place behind the Officers' Mess bungalow.

See page 114.

'I Mohammed Yamin, sahib. Your bearer, sahib. I have tonga for baggage. Today we are going to Green's Hotel;[1] tomorrow terrain, frontier mail to the paltan.'

Heavens, thought Littlewart, this is Kipling! 'Spose Kim will be doing tricks in the Servants' Class! He also thought, this feller is going to run me – not that I mind – but better watch out. In fact as time went by he found he could manage Yamin without difficulty but just now it was useful for him to run the show. Next day with Mohammed Yamin in charge the journey was child's play.[2] Littlewart and his bedding roll were installed in a 1st Class compartment (seat six sleep four) with a little shower room attached at one end and with Yamin in the servants' compartment alongside at the other. There were no corridors but from time to time Yamin would bob up with fruit or char and sticky buns from the pop-wallah and at the appropriate station escorted Littlewart up the line to the restaurant car for his lunch, dinner and breakfast.

At last they reached the station for the Ramrods and there on the platform to welcome him was the orderly officer. Whilst Yamin remained to deal with the baggage the latter conducted Littlewart in a tonga to the bungalow which was to be his home – or more correctly the small portion of it that would be.[3] The OO left him and in due course Yamin re-joined, to unpack a few necessities of life.

The bearer's arrival seemed to be the signal for a swarm of Indians to appear as if from thin air to surround and bombard him with high-pressure sales talk with hand and mouth. In no time at all Littlewart found himself measured for uniforms by the regimental dhurzee.[4]

'Blighty pattern no good sahib. Must having regimental pattern. Ramrods –' he metaphorically bared his head as he spoke the sacred word – 'Ramrods very regimental sahib, oh yes very.'

Of course the dhurzee was right, but he managed to wear out on rough shooting and similar activities the khaki drill items which the accredited tailor in England had assured him were correct. In a sense they were. They were not incorrect but just did not have that particular regimental cut and style. By the time the dhurzee had finished Littlewart discovered he had become the owner of a secondhand bicycle.

'This is most essential conveyance sahib' assured the salesman. It was too – the mess was a mile away and the barracks still further. The only other form of transport was Shanks' pony, so everyone owned a bicycle. Cars were very, very few and far between.

It seemed he had also engaged a munshi to teach him Urdu three days a week.

'To hold converse in native tongue' proclaimed that long white bearded dignitary, 'is most pleasant and desirable.' It was too – in fact everything that was not essential was most desirable. But he did fight the bannia who considered it quite essential to make him a loan at 'very small percents' to carry him on until his pay came through. 'Clanger' had heard too many tales of officers who had served two and more tours (over 12 years) and for whom India was an open prison because they could never scrape up the cash to bail themselves out of the bannia's clutches.

When at last the plague of locusts moved on it was past noon so Clanger decided to leave Yamin to continue with the unpacking and pedal down to the mess for a gin before lunch-er, tiffin! He was soon pedalling back. He had found the orderly officer having a bracer before returning to

1. Whilst the Taj Mahal had the bigger reputation, many old China Hands including Littlewart's bearer reckoned that Green's was more friendly and comfortable and certainly cheaper.

2. It was virtually impossible for a sahib to travel any but the shortest of journeys without a bearer. Apart from losing face there was a lot of hard work involved.

3. There was no living accommodation for officers in the mess building and unmarried officers lived chiefly in unwanted surplus accommodation in married officers' quarters. Occasionally there was a bungalow specially built for half a dozen single officers where the wilder types were liable to gravitate – the sort who were too lazy if they were in bed to put out the ceiling light by using the switch at the door but would shoot out the bulb with their revolvers. Such bungalows were commonly known as 'Cad's Alley' and it was quite an honour to be chosen to join this select (?) band when a vacancy arose.

4. Just as in England regimental tailors seem to get news of impending moves long before orderly room does. The first person to get the news of where a regiment was going to was the regimental contractor. He got his information from a clerk in Army HQ Simla or Delhi, on payment. He then informed the Colonel Sahib.

A tented Officers' Mess in India. circa 1905

barracks to do the men's dinners and he had advised him 'to cut back and get a jacket and tie on. Shirt sleeve order won't do *in* the mess you know'. Of course Littlewart did not know but he soon learned. And he learned that far from being more free and easy, a mess in India, at any rate the Ramrods, could be more regimental. At lunch he met many of the officers and some of them even spoke to him. Well, he did have his second pip up and a few years in; it would probably have been different if he had come straight from Sandhurst. Clanger smiled as he remembered the tale his father had told him of the member of the Athenaeum Club. He was approached by another member who asked him if he had finished with the paper on the table in front of him. At which he promptly burst into tears and begged excuse for this remarkable lapse by explaining he had been a member of the club for twenty two years and this was the first time anyone had spoken to him.

After lunch it was a quiet pedal back to the bungalow; off with the jacket and tie as a start and then a bit of Egyptian PT.[5]

* * * * * * * * *

His bearer brought him tea and toast in his bungalow after which he made himself known to the captain and his wife whose quarter the bungalow was; enjoyed a drink or two with them and soon found it was time to think of dressing for dinner. His mess kit was all laid out on the bed and Mohammed Yamin was running a bath for him; that is to say he was heating a large kerosene can of water over a wood fire in the compound! Empty kerosene cans were as universal and useful a commodity as expendable petrol cans were to be in the North Africa campaign. When the water was steaming it was poured into a tin tub in his bathroom, which also housed his 'thunder box', cold water was added and after conducting a search for scorpions or snakes his bearer announced 'Ghusal taiyar hai, sahib' (bath is ready).[6] The sahib took his bath and then dressed with the assistance of his bearer who would hand him his socks, his shirt, his tie, help him into his overalls and finally – the grand finale – wind him into his cummerbund, a made-up one was as unthinkable as a made-up evening tie. The drill is that the sahib holds one end of the cummerbund to his stomach, the bearer moves away until it is sufficiently unwound and then the sahib moves to the bearer rotating as he goes. Finally he is duly swaddled and the spare end is tucked neatly away. Simple enough, but an officer new to the technique has been known to contrive to get himself and his bearer inextricably entwined. Then it is hup! carefully mount the subaltern's charger – overalls are a bit tight at first and proceed gently down to the mess. And a few hours later return, perhaps not quite so steadily, by the light of the moon to the bungalow and bed.[7] Then in what appeared to be only a few minutes later, Mohammed Yamin would materialise out of the dawn

5. Or siesta, also known as charpoy bashing and working for the staff college.

6. Baths with running water and hot and cold taps would only be found in Residencies, or the homes of very high personages. They, like WCs that flushed, were unheard of in the homes of lowly mortals such as Army officers, married or single. To say that such a thing could exist on the frontier would stretch credulity too far but such was the claim of the Wali of Swat – a mountain State some hundred miles North East of the Khyber Pass.

During the last war he invited a Major at Malakand Fort to stay at his Palace. 'You Major Sahib and your Mem-Sahiba will then be able to enjoy the only long bath with hot and cold taps on the frontier. It is, Sahib Bahadur, the *only* long bath with hot and cold from the Hindu Kush to Baluchistan. It is the talk of the Peshawar bazaar and the wonder of the caravans from Cabul and envied by all other Nawabs on the frontier.

The Major and his wife arrived as the sun set behind the mountains. As the officer sat listening to sly digs at the Wali's ancestral enemy, the Nawab of Dir, the Mem-Sahiba went to her bath. She turned on the hot tap and piping hot water gushed out. She turned on the cold for a short while and then got into the bath. But it was now a little too cold. As she was about to turn on the hot tap again she had a feeling that she was being watched. She looked everywhere and finally saw a small hole above the taps through which an eye was peering. It was the eye of the ancient and trusted 'bhisti' (water carrier). He was not interested in the Mem-Sahiba but he had been given strict orders that when she turned on the hot tap he was to pour hot water into a funnel labelled HOT in persian script, and if she turned on the cold he was to pour cold water into the other funnel.

The Wali never knew that his secret had been found out!

7. Although the popular conception – at least on the music halls – of an officer in India was of a curry-gorging, burra peg swilling, mentally deficient, very probably as many pau pegs as burra pegs were drunk in the mess. The pau peg was a quarter tot and taken with plenty of soda was about as harmful as lemonade, particularly as it was sweated out as quickly as it was imbibed.

An Officer's bathroom or 'Ghusal Khana' in India.

The 'Bhisti' is carrying in two old kerosene oil tins containing hot and cold water for the Sahib's bath.

See page 116.

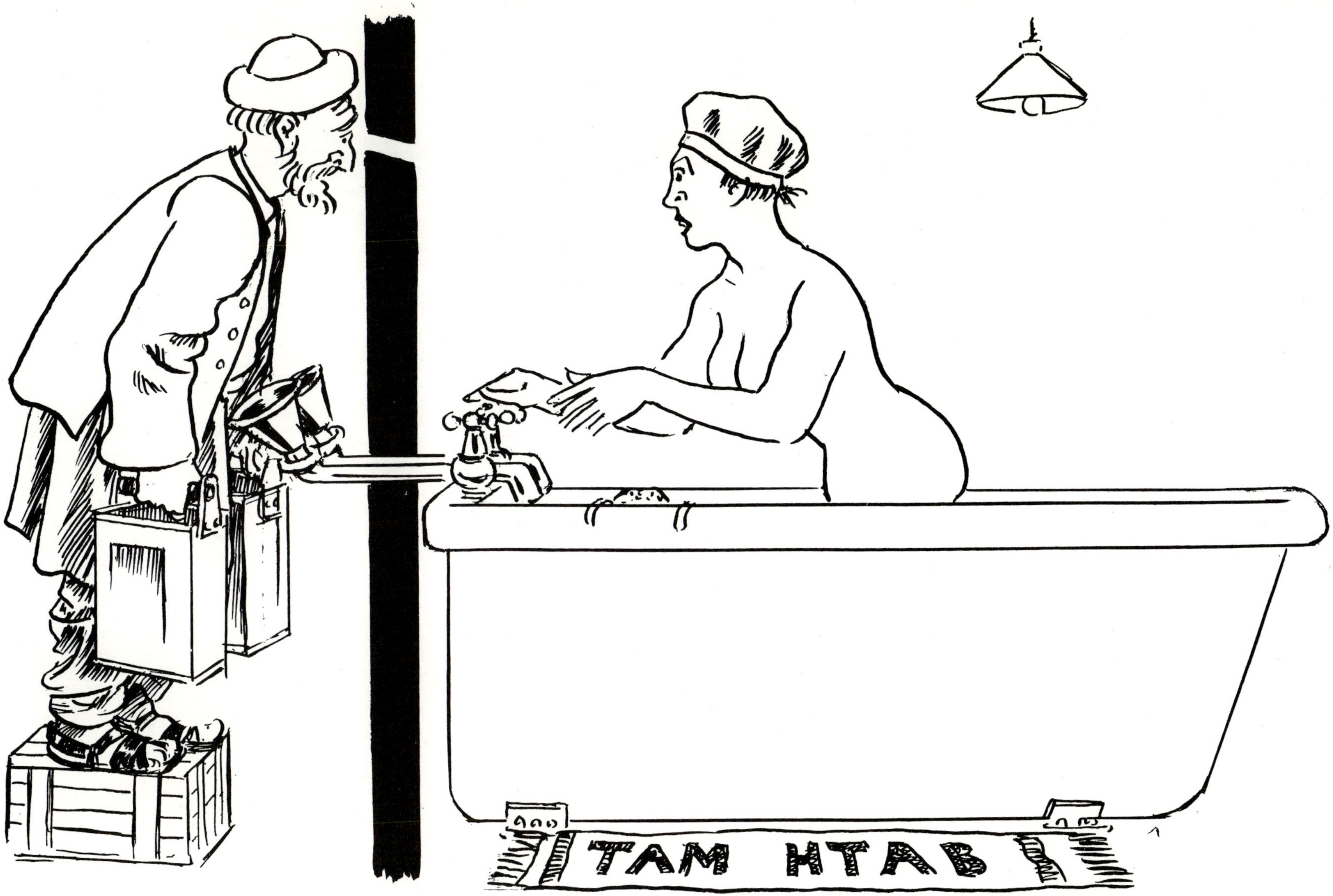

The latest 'innovation' – the Wali of Swat's new long bath with H & C 'piped' on.

with morning tea; and so it was going to be for some years to come because normally, unlike a batman in England, a bearer would stay with his officer throughout his tour in India. Another point of difference was that a bearer was almost certainly many years older than his master, in the case of a young officer old enough to be his father but nonetheless fit and tireless. Nevertheless, officer and bearer became very good friends in spite of the difference of race, tongue, and age. Such was the case between Jafir Ali a veteran bearer and his officer, who recalls – Jafir Ali was a Hindustani Mussulman, a native of the ancient kingdom of Oude. In World War I he had served with the 2nd Lancers and had taken part in Allenby's famous cavalry operations in Palestine. Like all old soldiers he liked to reminisce and his favourite tale always ended with 'Har ek admi teen char turki-log apna ballum per'. (Every man had three or four Turks on his lance.) But Jafir Ali never bored me; he was far too sensitive, too courteous, to chatter if one was not in the mood for it. And in all our hours of fascinating talk, while I lolled in a long chair under the punkah, he would be either standing, or, if it looked like a long bout he would squat on his hunkers. Sometimes he would even accept a cigarette, which he smoked through his fingers without ever touching the tobacco with his lips. It just never occurred to either of us that he should sit on a chair. There were no chairs in his own go-down and on the rare occasions that I visited it a chair had to be brought specially for me from my bungalow. In six years I never saw any of his womenfolk face to face.

The best way I can think of to describe my relationship with Jafir Ali is to tell of a typical day in our lives during the hot weather in the North of India. The day began with the sound of the pipes playing reveille and the ghostly figure of Jafir Ali outside the mosquito net bearing the tray of tea. He announced its arrival with the words 'Gun fire sahib'. Those were the only words of English I ever heard him utter. As I got up Jafir Ali was busy laying out the spotlessly laundered and starched khaki drill uniform with shining buttons and badges duly affixed. It crackled as one put it on, but on the way to barracks on a bicycle through already sweltering heat, the crackling quickly melted.

The next time to see Jafir Ali would be back in my bungalow again after breakfast in the mess. If there was a parade on that day or some duty requiring drill order – kilt, spats etc., – Jafir Ali always knew about it and had it all ready. I never discovered how he knew for he neither spoke nor read any English. When I asked him he just smiled his slow smile through his beard and replied 'Hum ko malum hai, sahib'. (I have knowledge, sahib.)

After lunch in the mess, where Jafir Ali would appear to serve me wearing white mess livery with tartan cummerbund and pagriband, I would go back to the bungalow. Jafir Ali always seemed to get there first, in time to announce the arrival of the munshi. Jafir Ali was inevitably around during my lessons doing quite unnecessary jobs about the house. When the munshi had left he was at great pains to correct the munshi's unacceptable Bengali syntax.

Soon after Jafir Ali came to me he appointed himself my master of horse and took complete charge of all my ponies, so if it was a polo day he would appear at about four o'clock riding out at the head of a string of ponies with the syces running along behind carrying sticks and all the other paraphernalia. By the time I got to the polo ground in the car everything would be as perfectly organised as a squadron of 2nd Lancers on annual inspection. Jafir Ali took a great personal interest in the polo game and was a very competent schooler of ponies. He supported the regimental team vociferously and I often wished that I had been a good player if only for his satisfaction.

Back in the bungalow, hot and tired, to find Jafir Ali always there ready to pull off my boots. Then he organised the bath water – which arrived in old kerosene tins on the bhisti's yoke – then my whisky and soda, with just the right amount of ice. As I lay in the bath I could hear Jafir Ali examining the ponies, trotting up any that he thought pottery – and fairly smartening up the syces. This after-bath-time was in many ways the best time of the day. One flopped about for as long as possible before dragging on absurd, tight, tartan overalls and white mess jacket. Now, also business was transacted. If one wanted some new clothes a little brown lad was sent running to summon the dhurzee, who came bowing and smiling with his tape-measure and his patterns, and if he did not come quickly enough Jafir Ali wanted to know the reason why.

At dinner in the mess Jafir Ali again appeared the perfect servitor behind my chair. He wore his medals at dinner – ribbons only at lunch. This was no mess rule – just Jafir Ali's idea of what was seemly. At nine o'clock the bearers were allowed to leave the mess and Jafir Ali could go home to his go-down in my compound where he lived with his family. He had cheerfully worked a fourteen hour day! This he did for 365 days a year on rupees 35 a month (about £3).

Next morning, without possibility of failure, Jafir Ali would be outside my mosquito-net at reveille with 'Gun-fire, sahib' and – if there was a suspicion of a hangover an Alka-Seltzer as well. How did he know? 'Hum ko malum hai, sahib'.

Most bearers were of the type of Jafir Ali but very occasionally one might go wrong as an officer who was stopping the night in the guest room of a mess on the North West Frontier discovered. He woke at 4 a.m. to find all the lights burning. A smiling Pathan was holding a large stone over his head which, if dropped, would undoubtedly have crushed his skull. Three other Pathans were quietly clearing the room of all his belongings.

The leader of the gang of thieves who had obviously once been a bearer, carefully placed a clean shirt, vest, shorts, and stockings neatly on a chair. Then with a courteous bow, smile and wave of the hand he explained 'For tomorrow Sahib bahadur'.

Then they departed, the boulder holder last.

* * * * * * * * *

'Going to Polo'
circa 1890.

Ordinary dinner nights did not differ very much from those in England, discounting the effects of the climate. A stiff collar might rapidly become soft, the soup might be diluted with the occasional drops of sweat from one's forehead when the weather was particularly hot and humid! But one got used to it and of course all dirty linen was washed and starched and returned beautifully pressed within twenty four hours for a dhobi's bill a fraction of an English laundry bill. So the days and nights went by until one day Clanger was invited to dine with the 1st Bn The Queen's Own Cameron Highlanders. Let one of his hosts describe the occasion in his own words.

'The best thing about a regimental guest night was that one never knew how it was going to be. It all depended on everybody's mood. If the chief guest was stuffy or the colonel was in a bad

The Guest of Honour, Governor of the Province, and ruler of millions of Indians, has 'Angels on Horseback' poured down his front by a new mess Khitmagar.

See page 113.

The Officers' Mess of an Indian Regiment on the line of march up the Grand Trunk Road.

See page 118.

A dramatic incident in a Mess on the West Coast of Africa.

See page 117.

temper, it could be four hours of acute boredom. But if the guests were young in heart and, best of all, the colonel was away, it could be a riot of fun.

One was first reminded by the orderly piper playing first pipes outside the mess. This told one that there was half an hour to change and usually emptied the ante-room, as it took longer to get dressed in the kilt for guest nights than in the every-night dress of strapped trews. Also on guest nights one was supposed to be in the ante-room five minutes before second pipes at 8 o'clock. If one had a private guest one prayed that he would arrive early so that one might introduce him to the CO before the arrival of the top brass. Second pipes had to wait until all the regimental guests had arrived.

Then began the calls of "koi hai"[8] which brought the bearers hurrying silently on bare feet in their flowing white garments, with Cameron tartan cummerbunds and pugari-bands which were also adorned with a large brass '79'.

While everyone was being refreshed the senior subaltern went round and detailed four subalterns to dance the set *Reel o' Tulloch* after dinner. Anyone might be detailed, so that everyone had to know his stuff. The only man excused would be the orderly officer, who by 10 o'clock would have departed on his bicycle with his sword clanking on his handlebars to take Last Post at the Guard room.

When everyone was refreshed the mess sergeant would appear in the doorway and shout above the din, "Gentlemen, dinner is served". Then with the colonel leading the senior regimental guest, who would sit on his right, everyone crowded into the dining-room.

The scene in the dining room never failed to delight me. The long mahogany table was loaded with silver of the most flamboyant style. It was dominated by a figure of winged Victory on a plinth about four feet high. The plinth was surrounded by four two-foot-high figures of Camerons of the 1870s. Along the table was ranged more elaborate Victorian silver-ware and a candelabra. In the dim light, outside the glow of the candles on the table, hung at one end the King's and Regimental Colours. At the sides was a collection of flags and banners captured in battle, including the Queen's Colour of the 38th Native Infantry, which had been carried into action against the 79th Highlanders by the mutineers in 1857.

Throughout the meal, which was the usual five course affair, the military band entertained us from the garden. When the table had been cleared and all wine glasses charged, the mess president at one end of the table rose to address his colleague at the other end; "Mr.Vice, the King Emperor". Amid the clatter of chairs as everyone rose, Mr.Vice called out "Gentlemen the King Emperor"[9]. At that moment the mess sergeant made a signal to the band who struck up the National Anthem, while we stood clutching our glasses. Then amid murmurs of "The King, God bless him" we emptied our glasses and sat down. While the coffee, cigars and snuff were circulating the bandmaster left the band and came and sat beside the colonel who poured him a glass of port.

8. No pressing the tit in India. Waiter service was obtained by calling 'Koi-Hai', i.e. 'any body there'. The answer 'Sahib' was from a Khitmagar usually listening to the Sahibs' brilliant conversation round the corner sitting on his haunches. As a result of this old soldiers from India were known as 'Koi-Hais'.

9. British regiments in India and Indian Army regiments toasted 'the King Emperor'.

Painting by Col. Monro.

1st Bn. The Queen's Own Cameron Highlanders.
A cold weather evening in the Officers' Mess in Fyzabad United Provinces, 1932.

Captain J. W. Fraser

Lieut. D. Macfie
Major H. MacLeod, RAMC
Lieut. A. E. H. Campbell
Major R. A. Colville, MC.
Captain C. A. MacLeod

Lieut. W. MacDougald
Captain C. S. MacNab
Lieut. I. P. Grant
Lieut. B. D. Hope-Dunbar
Lieut. M. G. O. Hill, adj.

Lieut. A. J. J. Somerville-McAlester

'Officers' Mess – Second Call'

Now came the *piece de resistance,* the entry of five pipers in full dress, their kilts swinging, their accoutrements flashing and their pipe banners swaying to the rhythm of their march as they pounded round the table.[9] After the march the pipers formed up at one end of the room and played a selection, eventually leaving to another march tune. After an interval for filling glasses the pipe major came in alone playing the pibroch. Sometimes a guest unaware of the solemnity of the moment went on talking but the dirty looks soon made the poor man subside.[10] At the end of the pibroch the pipe major made his way to the colonel's chair where the mess sergeant waited with a bottle of whisky and a silver quaich. The colonel rose and poured a dram into the quaich which the pipe major took, proposed a toast in Gaelic and poured the dram down his throat.

Next came another set from the pipers, but this time when they formed up for the strathspey and reel, the four subalterns previously detailed got up and took position for the Reel o' Tulloch.[11] The CO watched every step so one was very careful not to put a foot wrong. When the dance was over one returned panting and sweating to one's place and the pipers marched away to the *Pibroch o' Donuil Dubh.*

At this point the bandmaster, refreshed by the colonel's port, returned to his band in the garden. Their finale was *The March of the Cameron Men* sung unaccompanied with the officers joining in. All those male voices singing the old tune was very moving, though one would not have admitted it at the time.

This ended the dinner party proper and thereafter it was free for all; billiard fives in the billiards room; childish games on the ante-room floor; or more general and rougher reel dancing to the one piper left behind for the purpose – and he has been known to stay until reveille. But usually with the last official guest the colonel departed and the married officers drove off, sucking peppermints, to their wives. The younger chaps, according to their moods eventually dispersed to their bungalow. Sometimes on the way tonga races were staged with the wretched tonga ponies whipped to full gallop by their kilted drivers, while their rightful owners begged for baksheesh in compensation.'[12]

* * * * * * * * *

Before leaving India after this very brief tour, mention must be made of a character 'and character he was indeed' a member of mess staff unique to India, the Abdar. The vivid and obviously heartfelt description of this man of many parts which follows has been contributed by Major General P. Gleadell CB CBE DSO, late of the Devonshire Regiment.

'The Abdar was monarch of the nether regions of the mess compound and Keeper of Custom in the ante-room. Exercising control on behalf of the PMC and mess sergeant over a host of Khidmutgars, Khansamahs, Bhisties, Sweepers and Officers' Bearers (and all their families) his knowledge of informed gossip was little short of miraculous.

Barefoot, rotund yet deft of movement, Abdul Rahman would appear silently from nowhere at the clap of a hand at any time of day and night – whether at the behest of the orderly officer

9. Sometimes a Hazard attended this perambulation. Once when the 2nd Bn The Seaforth Highlanders were dining in India the tassels of the pipes of a nervous young piper got caught in the mantelpiece as he went by and they 'died, swinging and squealing while all the pipers marched round once more before leaving the room'. It says a lot for the respect of a Scot for the pipes that no-one laughed until all the pipers were safely out of the room.

10. Needless to say Clanger Littlewart transgressed in this way. He was reaching the climax of a particularly risqué story he was telling his neighbour and thought that the growing silence was due to respect to his gifts as a raconteur!

11. Or in the Seaforths the CO might call on the pipe major to return at this and play the 78th 'Walk Round'. This would be the signal for the junior officer present, followed by the officers next to him, to form a chain behind him, his own hands being on the pipe major's shoulders. The pipe major would then lead the crocodile of officers out of the room. A good way of clearing the dining room but of course not to be adopted until a suitable moment had been reached.

12. In the hot weather, if you could get hold of a car, a favoured after dinner-night entertainment was to drive to some unfortunates bungalow and see how far you could push his charpoy – he would be sleeping out of doors – across the compound before the whole thing collapsed in a tangle of occupant and mosquito net.

A Scots Regiment in India – the Orderly Officer's vigil scene in the Mess ante-room at midnight. After a hectic night, followed by a long day on duty, the Orderly Officer is worn out.

See page 117.

returning from late night "rounds", or to satisfy the needs at dawn of officers returning from manoeuvres. Unruffled, always impeccably rigged in regimental cummerbund and pugaree, the unforeseen bore him no problems. Just once, perhaps, he allowed his reserve to sway. The occasion was the return of the orderly officer, at the close of a guest night, having taken Staff Parade of the orderly sergeants. Abdar greeted him in almost jovial mood saying "Dekho, Dadson sahib pantaloon ke punkah". And sure enough, there in the ante-room was the very widest subaltern in scarlet jacket, but trouserless, his overalls waving madly round on the blades of the huge ceiling fan. But it was never revealed that Abdar, in hasty consultation with the mistri (electrician) had engineered a fault in the switch which kept the punkah in perpetual motion.

A tactful recipient of telephone messages, he was ever a discreet reminder of dates, perhaps overlooked in the dinner Warning-out Book. Young officers accorded him the sort of deference they did the RSM or the bandmaster and the affection they might for a kindly but rather critical uncle.

He spoke no English – but understood every word of it. He knew without reference to the inventory every detail about the mess silver, and he would almost genuflect when he uncased the Colours in the mess before a guest night. Serving a late round of drinks one night a guest said to him "Surely, Abdar, you would agree with me that tradition is the voice of error echoing down through the ages. Take Balaclava, Kanpur or the retreat from Mons?". To which Abdul Rahman replied, "Ne sahib. Tradition bahut acha bandobast".

In common with most Indian servants Abdul kept a book of chits or reference from previous employers, of which he was immensely proud and they included those of a forebear who had been Abdar in the Residency in Lucknow at the time of the Mutiny. To be shown this book was a privilege and he who was allowed access could rest assured that he had been considered to have reached the standards of acceptability.

But his *tour de force* came at the passing of the port at a guest night when at opposite sides of the dinner table he and the mess sergeant would each take hold of one of the long narrow strips of table-cloth and in one flick whip them away to the applause of the diners and cries of "Shabash Abdar!".

Of course there have been bad Abdars as there have been bad Popes and so a mess temporarily slipped in esteem. But these were rare and Abdul Rahman was one who tended well his very divers flock. That servants like him were reared over the years is perhaps not a bad indication of what the officers' mess really meant.

And, dare one add, the British Raj too?'

'Hunt the Po'

The pack follow the 'scent' in this exciting pastime, which was carried out by the subalterns of the Royal Scots Fusiliers after Guest Nights.

See page 116.

GLOSSARY OF URDU WORDS

The following is a short list of some of the words that Littlewart would have found in general use around the officers' mess and quarters. The spelling of some will without doubt be open to correction – but no two people seem to spell Urdu words alike. The spelling used, therefore, is a popular version but not necessarily the authorised. Where thought desirable phonetic pronunciation has also been given.

Abdar – Major-domo of the officers' mess. A butler, more properly 'a water cooler'.

Bucksheesh – Tip, gratuity.

Bahadur – A hero or champion. Used after the names of European officers (bestowed by the Moguls).

Bandobast *(Bundobust)* – Arrangement. (Bandobast Karna – Get everything ready.)

Bannia - Bania - Baniya – To rhyme with 'done yer'. A trader in grain, a merchant, shopkeeper or moneylender. Always a Hindu. He would relieve you, with a most charming smile of your rupees as surely and thoroughly as the mendicant would annas.

Bearer – Strangely enough there appears to be no Indian word for this Jeeves of the British Raj.

Bhistie - Bihishtie *(Beastie)* – Gunga Din – the water carrier, mostly employed in the mess and officers' gardens, brought water for baths. The water was carried on the back in a mussuck, or goat skin collected from a well.

Blighty – From Blayut, Billait or Wallyte – from the Arabic 'Wilayat', a kingdom or province. From this came the word 'Blighty' used by soldiers to refer to Great Britain and Ireland.

Bolo the Bat – Speak the language.

Burra – Big. (Burra Sahib - important Sahib, Burra Peg - large whisky.)

Chae – Tea. 'Char' to the British soldier.

Chapatti – a flat circular disc of unleavened bread.

Charpoy – An Indian bed.

Cummerbund – About three yards of silk in the regimental colour wound round the officer's 'massif central' instead of a waistcoat. Indian cavalry used the lungi or turban of a Sowar (trooper).

Daffadar - Duffadar – Sergeant of Indian cavalry.

Dekho – Look. 'Dekho sahib, tonga-walla has been struck by thundering clap'.

Dhurzee - Dhirzi – Tailor.

Dhobi – Laundryman. The 'Flying dhobi' was not airborne he merely offered express service if twenty-four hours was too long a period!

Jaldi – Quickly.

Khansamah – Cook.

Khidmutgar - Khitmagar – Waiter.

Kitna – How much?

Kush - Kushi – Pleased, happy, easy when referring to a person.

Makkhan – Butter

Munshi – Teacher, considerably more qualified than the babu 'failed BA'. Taught Hindustani to British officers.

Paltan *(Pultarn)* – Regiment of Indian infantry, from the French 'Peloton' and later the English 'Platoon'.

Pani - Pawnee – Water.

Pathan – A tribesman of the NW Frontier, known to the troops as 'They Paathans'.

Pugaree *(Pugari or Pagri)* – A 'bandeau' in regimental colour which was worn diagonally around the head-dress by bearers and mess servants. Also a turban.

Punkah – The big ceiling fan with blades about two feet long which circulated aloft, usually rather sluggishly, like some great albatross somnolently patrolling the Sargasso Sea.

In the hot weather when 'le cafard' struck the punkah has been known to be turned full on whilst subalterns danced beneath it throwing old unwanted gramophone records up into it. This was quite a lethal pastime and it would have been wise to wear a Battle Bowler or at least a Bombay Bowler; but the fun of course lay in not doing so!

Qu'hai - Koi-hai – Or however it is spelt. The meaning is roughly 'Anyone there?'. It is used to summon Abdar or a Khitmagar from the nether regions of the mess.

Roti *(Rooty)* – Bread – from this came the 'Rooty Gong' i.e. medal for long service (and undetected crime).

Shabash *(Sharbarsh)* – Jolly good; well done.

Sowar – Indian cavalry trooper.

Sweeper – His Indian name was Meta but he was invariably called sweeper. The soldiers, pronouncing the word to rhyme with metre, used it as a term of endearment 'You old meta you!' There was nothing endearing about his job which was emptying the sahib's thunderbox.

Syce – Groom.

Thunderbox – No exact Indian equivalent. English equivalent 'Elsan chemical closet'; except that this contained no chemicals. No wonder sweeper was the lowest caste in the system. However, he had a degree of protocol and his customers were graded as 'burra fujjer' (the dawn serenader.....'the dawn came up like thunder from China 'crost the bay'); 'hazri bad' (after breakfast); 'kubbi kubbi' (irregular); and 'kubbi neh' (corked – due for a number nine from the MO).

Tiffin – Lunch.

Tonga – A one horse cab; sometimes also called a gharry; a taxi irrespective of horse power was a motor gharry.

Topee – Pith helmet worn by Englishmen to protect themselves from the mid-day sun. Not to be confused with the Wolseley Helmet – an acutely uncomfortable head-dress worn on ceremonial parades.

It was also generally worn off parade by the officers to distinguish themselves from the box-wallahs who wore a lighter and smaller model known as the Bombay Bowler.

Tum-Tum – Dog cart driven by officers before the days of motor cars.

Wallah – Man. Usually used with another noun to denote trade or profession. Hence box-wallah, anyone engaged in any form of commerce; pop-wallah the lemonade seller; the familiar char-wallah or tea-purveyor; tonga-wallah cab driver and loose-wallah – a troops name for a thief, probably a rifle thief from the frontier. Used with a verb it can denote a physical state. Thus to be a 'Frontier going chap', merely meant that the person was going to whatever the destination quoted. One could also be a sleeping chap or eating chap!

viii

Mess stories - all short, none too tall

An unfortunate motoring accident in England brought Littlewart, like the ill wind, nothing but good. A senior major had been killed and a captain crippled and forced to transfer to the Pay Corps. So two vacancies for promotion to captain were created and to his surprise Littlewart found that he was to fill the junior vacancy. Thus in the Ramrods he contrived to get his captaincy in eleven years, much to the disgust of the senior subaltern of the regiment alongside, who had already got in half as much again.

The vacancy existed in the home battalion. Like all other regiments the foreign battalion of the Ramrods was kept up to full strength and owing to the vagaries of the trooping season[1] usually a bit over, so although he had not yet completed his foreign tour of six years he reverted to home establishment. He returned to England with a healthy bank balance and a posting to the regimental depot where he found life very pleasant and peaceful and, being of an age when he would not 'live in sin' if he married, he did that thing.[2] Then came Munich, ARP, and digging trenches in the mess garden and finally the 'real thing'. Littlewart found himself winkled out of the depot and, as a founder member of the BEF plunged into a war that was far from the real thing. It would not have done for Errol Flynn at all! He survived Dunkirk and went airborne; got fairly badly wounded at the Sicily invasion and found himself faced with a long spell of inaction. So he decided to write his memoirs. Well, he had to do *something*. He was not a successful escaper from Colditz, he only wanted to escape from the intolerable boredom of the British Military Hospital. All he could think of, however, were the many funny situations in which he had been or of which he had heard about other officers. He identified himself with these situations whether or not he had been present and recorded them in the hope that they might provide a basis for his story. There are not many because when he was fit enough to be returned to duty he was sent to the Staff College to improve his health and his military qualifications. He did quite well, thus, as he said, proving the Staff College motto 'Tam tauro quam cerebro'.[3] During and after the Staff College he had no time to indulge his literary fancies but the tales he collected remained. Here they are.

* * * * * * * * *

Live-stock in barracks[4]

The keeping of livestock in barracks has always been subject to strict control by the 'Regulations' – as was so well pointed out by Anthony Armstrong, that splendid humourist who painted such

1 The trooping season in those days was limited generally to the cold weather. Thus a young officer posted to the foreign battalion in India in May might wait until the following September before he actually embarked on his journey. Similarly an officer due home in July might have to wait until October before he got on the move. It was not, like now, an all the year round process.

2 An officer was eligible for marriage allowance at the age of thirty. If he chose to marry before then he got nothing and was popularly described as living in sin, though as far as the Church and the law were concerned quite legitimately married.

3 The Staff College motto is Tam marte quam Minerva. Littlewart's version was quite popular during the war. (Sometimes translated as 'Too many Gunners get on my nerves'.)

4 In fact the equestrian was Major John Onslow of the DCLI.

88th Foot - Officers' Mess, Tralee, Co. Kerry. circa 1846

Major Bayley's 'Light of Love' jumped the mess table, a good four feet and a half without laying an iron on it, and landing safe on the other side. He stood, gentle and quiet upon the floor under which lay a wine cellar of from ten to twelve feet deep.

happy pictures of army life soon after the Kaiser's war. The situations he depicted embracing such animals as monkeys, hens and tortoises illustrate how wise it was to lay down strict controls — and hope for the best!

As far as the officers' mess was concerned, livestock was normally represented solely by dogs. Budgies were not favoured, nor were many goldfish in ponds in evidence and although many officers owned hunters and polo ponies it was not the normal practice to introduce them into the Mess, though it was not unknown. Rudyard Kipling, in his magnificent fictional short story *The Maltese Cat* — as stirring an account of a sporting event, fictitious or factual, as probably has ever been written — describes how after the epic match between the Skidars and the Archangels that 'Past Pluperfect Prestissimo Player of the Game', the Maltese Cat, entered the dining room after dinner where 'he behaved himself very politely.....and they drank his health'.

This might nowadays be considered a tall story even for Kipling's army of the 90s but it is not by any means. In the RA of the thirties ladies were not allowed in the mess proper; there were ladies rooms where they could be entertained on occasion. Horses on the other hand (reports a Gunner) were frequently invited to enter the mess — e.g. well behaved polo ponies who had distinguished themselves in a tournament, or a particularly handy hunter who could be trusted to jump over the furniture on a guest night.

In the officers' mess of the 88th Foot (The Connaught Rangers) Major Bayley's 'Light of Love' jumped the mess table — a good four feet and a half — without laying an iron on it and, landing safe on the other side, stood gentle and quiet upon the floor under which lay a wine cellar of some 10 to 12 feet deep. It is not reported whether he took wine with the members. But it is possible to report truthfully an occasion when a horse took tea in the ante-room.

Without unduly disturbing a somewhat somnolent duty groom I had extracted my little horse which I was allowed to keep amongst the mounted officers' chargers. He was a pet and to my mind was more intelligent than any horse with which I had come into contact. He was balanced and fearless. I was not accoutred for equitation, nor was my mount, I merely wanted him for company. I had not bothered to bridle him and my only means of control was a rope attached to his head collar. I was passing the massive mess building when the idea came to me to try the horse under unusual conditions. We entered the mess by a side door and clip-clopped noisily down the stone floored passage. The horse paused once and with pricked ears inspected the bathroom. At the end of the passage were double swung doors which gave access to the main entrance hall. Luckily these were open and we were on a carpet which gave a less precarious foothold. The door to the ante-room was open and we progressed fearlessly between tables, armchairs and other furniture. Asleep on a settee was Tom, a crumpled newspaper on the floor by his side. My horse lowered his head to the cucumber sandwiches and other delicacies on the tea table and scoffed the lot. The noise of clinking china woke Tom. He sat up and surveyed us. There was no surprise on his face. 'He's eaten my sandwiches and my bun', said Tom reproachfully. 'Clanger old boy would you ask your bloody horse to ring the bell?'

* * * * * * * * *

Scarcely a dog's life

The Indian Infantry were understandably jealous of the Indian Cavalry who were stationed in the best stations, lived very happily beyond their means, and due to being seen playing polo, out shooting and pigsticking, and at social functions in lovely uniforms usually got plum jobs on the staffs of the mighty from H.E. the Viceroy downwards. The officers of the Indian Infantry on the other hand sat in forts or perimeter camps surrounded by wire, on the frontier, where men were men and women were not. Not unmindful of this the Colonel of a famous Punjabi regiment in Ferozepore called his cat 'Indian Cavalry' and when a shy young officer of a British regiment asked him why, he explained, 'Well me boy, y'see, when he is not eating or sleeping he is playing games, drinking or making love'.

* * * * * * * * *

Caught on the rebound[5]

In his time Clanger had been engaged in many an after dinner rag on a guest night but seldom had one rebounded on him as this one did.

A guest night coincided with that of a Highland Regiment which lay alongside us in Marlborough lines, Aldershot. Above the tinkle of our civilised band I could hear the wail of the Highlanders' pipes and before we had reached the stage of Angels on Horseback (a stable course for dinner) evil thoughts permeated my mind. The enemy, amongst whom we had many friends, would be up to their highland games after dinner; they would be defenceless; we would strike in secrecy. We planned the operation in the billiards room and then proceeded to execute it.

We uprooted every notice board in the vicinity and privily put them through the open windows of the officers' bedrooms. I remember that my board had on it inscribed 'To the NAAFI'. We awaited events and prepared to repel boarders but nothing happened.

Next morning the senior subaltern of the Highlanders payed us a courtesy call. He was worried, he said, as he sipped his sherry. (It was Saturday and the half day's work was done.) We were solicitous. Could we help? 'Somebody', said the large Scot 'put a NAAFI notice board on the colonel's bed last night and he is raising hell. He's out for blood'.

'Any good if I come and apologise?' I asked.

'I don't think so, it's too late', said the Highland officer lugubriously, 'but I suppose you could try.'

5 It was also John Onslow who was the victim of this leg-pull.

I followed him to his mess. The faces in the ante-room were sad but friendly. 'The CO's in his quarters', said someone. I was led there. 'Best of luck!', whispered my guide and tapped on the door. A voice roared 'Come in' and I found in front of me in a room devoid of furniture a row of kilted subalterns. I think they won.

Their commanding officer was on leave!

* * * * * * * * *

Off on a false scent[6]

It was a custom at the Depot of the Duke of Cornwall's Light Infantry at Bodmin during the mid-thirties to invite, on occasions, the North Cornwall Hunt to hold a lawn Meet at the officers' mess. Such an invitation (which included Littlewart who was on leave at the time) was accepted for a day in January 1934 during my tour of duty at the depot. I was PMC at the time and on that particular morning I was making the necessary arrangements for the sustenance of those likely to attend the Meet, when the Master of the Hunt telephoned to say that owing to hard frost the previous night, the Meet must, regretfully, be postponed to a later date.

Certain arrangements had previously been made for the reception of the visitors at the main gate to the barracks so, as I was orderly officer that day, I took the opportunity of instructing the orderly sergeant, whom I was due to meet in that area for our joint inspection of the meat-store, to inform the guard commander and any other persons concerned, of the altered situation.

My rounds as orderly officer completed, I was partaking of a well earned restorative in the ante-room when the mess waiter informed me that the RQMS wished to speak to me. With a grave face the latter then asserted that he had been informed by the corporal in charge of the meat store that I had pronounced that the meat due to be issued to the cook sergeant and which I had previously inspected on my rounds, was unfit for human consumption. The RQMS and I hastened to the meat store where the adjutant and the quartermaster had already arrived on the scene, having somehow got wind of the apparent crisis.

A post-mortem ensued at which the quartermaster, having inspected the corpses, announced firmly that they showed no signs of putrefaction. Enquiries were set in motion to find the source of this unfounded rumour. It was eventually traced to the orderly sergeant who had misunderstood the instructions I had given him earlier in the morning, which had been 'Please inform anyone concerned that the Meet's off'!

* * * * * * * * *

Undress Uniform[7]

In 1956 Colonel Littlewart DSO, who was coming to the end of an appointment as a Colonel Q at HQ BAOR, was a guest at a regimental guest night of the 1st Bn The Devonshire Regiment at Celle. He was only in the supporting cast, the guests of honour being some officers of the famous Canadian Regiment – the 'Vingt-Douze'. After dinner the fun naturally became rather fast. There were few pompous, male 'blue stockings' in the Canadian army. At least that had been Clanger's experience – at any rate *after* dinner. He and the CO of the battalion were chatting with one of the Canadian officers when the fun began to get furious and a frenzy of de-bagging set in. The officer could not help noticing this and also that he was about the only Canadian officer left still fully dressed. So putting his drink down on a side table he proceeded to debag himself, neatly folded his overalls and hung them over his crooked left arm; and took up his drink and the thread of his conversation with the CO where he had left off; all with the greatest *sang froid*. In fact if anyone was embarrassed it was Littlewart and the CO who feared they might be expected to follow suit!

* * * * * * * * *

6 The meat was condemned by Major D.Y. Hext also of the DCLI.

7 This tale is told by Lt.Colonel J.K. Windeatt OBE of the Devonshire Regiment.

Clanger was having his pre-dinner aperitif with some of the subalterns one night when the senior subaltern entered and said, 'Look you chaps I've just seen George on his way down here without a black tie on. Don't any of you say a word to him about it'. Senior subalterns had a way with them in those days and we all filed in to dinner without anyone having said anything to the normally impeccably-dressed George. Halfway through dinner the mess sergeant, resplendent in full evening dress and carrying a small silver salver, advanced to where an unsuspecting George was sitting talking to the adjutant.

'Excuse me sir,' said the mess sergeant, proffering the salver to George. In the pregnant hush that had fallen upon us all George looked at the salver. Upon it neatly folded, lay a double-ended black bow tie.[8]

* * * * * * * * *

The height of dignity[9]

King Christian of Denmark stood some 6ft.7in. in his stockinged feet. One day when visiting the 2nd Bn The Buffs Royal East Kent Regiment – he was Colonel-in-Chief of the Regiment – he entered the officers' mess and hung his service dress cap on one of the many antlers that decorated the upper part of the walls of the entrance hall. When an hour or so later the time came for His Majesty to leave there was a rush of subalterns to collect the royal Sam Browne, sword, greatcoat and other accoutrements. All were handed to him with the exception of the royal cap, which even the tallest officer standing on tip-toe could not reach. With a twinkle of genuine amusement in his eye, and to the delight of all those present, the King reached up and without difficulty retrieved his headgear.

* * * * * * * * *

Snakes alive[10]

Whilst he was in India, Clanger had on one occasion to share a bungalow with the CO. He could not remember exactly why but the fact that he did he was not allowed to forget. Returning late after dinner in the mess he was alarmed to see a large snake under the CO's bed on the verandah. He warned the CO who was awake, fetched his shot gun and blazed off. When the dust had cleared there was no sign of the snake. The shot must have missed but the snake had decided on discretion and had made good its escape. Next morning the CO was held up in his dressing. He could not find his socks. Yes, Clanger had scored a bull all right! The socks were eventually found in a nearby bush liberally peppered with pellet holes.

* * * * * * * * *

Boots

From socks to boots is but a short step but the boots of this story recalled by an officer of the Queen's Own Royal West Kent Regiment, have more affinity to 'buttons' than leather.

Before and up to the first world war there had been an old custom that it was the duty of the junior officer in the ante-room at the time to mark any game of billiards in which a captain or above took part after dinner. The signal was a loud cry of 'Boots' whereupon the young officer did his stuff. This was re-introduced after the war but subalterns then had a more independent attitude (or were not so well disciplined?) and they decided to kill the custom once and for all.

So it was arranged that on a certain night after dinner all the subalterns – less the two senior subalterns, who had not been let into the secret – would leave the mess quietly and go to their quarters.

8 This contravention of dress regulations occurred in the Buffs and is recalled by Colonel H.R. Grace OBE, DL, JP.

9 For this also we are indebted to Colonel Grace.

10 This happened to a CO of the Essex Regiment and is recalled by Major T.R. Stead DL, the Essex Regiment.

Sure enough a game was made up and the cry of 'Boots' came. The second senior subaltern, the junior of the only two in the mess, found that he was called upon to mark. He was not amused.

Fortunately the commanding officer had a sense of humour, considered that the junior subalterns had made their point and quashed the custom.

* * * * * * * * *

Messing and Mating Matters[11]

In a mess in India with which Clanger was very familiar the members had every reason to complain of the soup. The cook was an Indian Christian called Joseph who was always getting drunk – though an excellent cook. One night his waiter, Anthony, came in to serve the soup with his mouth bleeding. When asked what was the trouble, Anthony, a man of few words but liking to air what English he had, explained briefly 'Joseph hitting'. The next night he recommended that the officers should not take the soup. When asked to explain why not he merely replied 'Joseph spitting'.

One can only imagine that drunk or sober Joseph this time got the sack!

Eggs in India were, and no doubt still are, very small and until 'moorgi', that dead certainty for all restaurant cars and station hotels, can change its bore will ever remain so. At one breakfast[12] the khansamah complained that Littlewart sahib had eaten twelve eggs. (And where was his profit going to go if this went on.) Littlewart sahib could only explain that he had eaten the first four and was still starving; that the next four had been bad; and that he had eaten the last four. But Littlewart himself did not complain. He knew only too well that too many complaints would merely put him straight on to the mess committee as messing officer for the next quarter as had happened to an officer who constantly complained of stomach pains. This did nothing to alleviate these pains – if they existed at all – but at least it did lessen the suffering of his fellow officers who had to listen to his constant belly-aching!

The approach to complaints is fairly illustrated by some entries in the complaint book of the 1st Bn The Dorset Regiment.

Complaint The porridge at breakfast has not been good lately. This morning it was partly cooked and quite uneatable.

Complaint I entirely disagree with the above remark about the porridge being undercooked. The trouble is that it is always overcooked.

What does the messing officer do?

Answer I suggest a Board composed of these signatories should sit and decide what is cooked, uncooked or overcooked porridge. When the decision is reached it will be possible to give concrete (nice word) direction to the mess Khansamah.

The strange thing is that these entries were made not when the battalion was at Catterick but in India. No wonder the Dorset Regiment's motto is *Primus in India.* On this I would rate them as *Suprimus in Mundo.*

And from the complaint book of The Royal Scots Fusiliers, c1870. Any port in a storm, but let's have good port in the mess.

At the depot on one occasion the depot commander was lunching in (as was Littlewart) and he expressed in no mean terms his displeasure with the soup produced by the then 'lady cook'.

11 Major M.H. Bates, late Gurkhas contributed this culinary gem.

12 'Soldier' Thompson-Royds of the 2nd Royal Scots Fusiliers was the glutton for punishment in the egg eating contest, not Littlewart.

Unfortunately the doors to the kitchen were open and as the unfinished soup was returned by the waiter with the major's comments the said lady was heard to say, 'You may tell the major that he is no gentleman, too'. Which, truth to tell, Littlewart also thought but was not so well placed as the cook to say![13]

* * * * * * * * *

A lunch with which no fault could be found was given in 1940 by the chief of the French liaison staff to some senior officers of the British Army, but the Quartermaster-General Lt. General Sir W.G. Lindsell felt bound to comment. 'What a marvellous meal, but how can the French army work in the afternoons after such good meals?'. The proof of that particular pudding was unhappily in the retreating.

But whilst on the subject of work there is the tale of the two officers of the Irish Guards[14] who were relaxing in the ante-room over a glass of port in the mess at Windsor and, regrettably, talking shop.

Said one to the other 'You know Desmond, I think that all this business of sex is overrated'.

'Do you mean not too good for the poor?'

'No no, not that, just overrated. As that Jock said "The poseetion is rideeculous, the pleasure momentary and the expense prodeeegious". Too little return for the effort I say.'

'Oh, I wouldn't say that Major – haven't found it so myself, but of course remember, when the mind makes a contract the flesh can't fulfil, you're over the hill boy – er – Major you're over the hill.'

'I am *not* over the hill Captain FitzGerald. I just claim that it is 80 per cent hard work for 20 per cent pleasure.'

'Well, I would say it is exactly the reverse.'

Naturally as Irishmen, they could not agree but just then Guardsman O'Toole passed by the window on gardening fatigue and they called him in to settle the dispute. He was questioned by Captain FitzGerald.

'O'Toole d'you know what sex is?'

'Sorr?' (inquisitively).

'I mean do you understand what is meant by sexual intercourse.'

'Sorr' (emphatically).

'Well, Boyne here says it is 80 per cent hard work and 20 per cent pleasure, but I say it is exactly the reverse. We will accept your decision. What do you say it is?'

13 This occurred in the Essex Regiment Depot Mess, not the Ramrods.

The 'lady cook' can have been none other than Mrs. Purser, wife of Sgt.Purser who for so many years was mess sergeant to battalions of the Essex Regiment. She must be able to claim to be the first and last female member of an officers' mess staff – before the days of ATS and WRAC – to serve overseas on semi-active service. In 1934/5 the 1st Bn The Essex Regiment were part of the Force sent to the Saar to see fair play in the plebiscite. It might be termed the first United Nations Force although then, of course, it was the League of Nations and no gongs were issued. Many felt it was merely a curtain raiser to the next war. Nevertheless it was thought impossible that the battalion could go without their mess cook so Mrs. Purser duly accompanied the mess to the Saar as a 'tourist'. As a pair, the Pursers were incomparable and invincible.

14 I hope my friends in the mess of the Irish Guards will forgive me for tying this tale round their necks. No soldiers in the world can conjure so much nuance and expression out of the one syllable SIR as can Guardsmen, and none better than Irish Guardsmen.

'Sorr' (hesitantly).

'Yes?'

'Sorr permission to speak?'

'Of course.'

'Sorr. Well, I'm after thinking it must be 100 per cent pleasure.'

'How do you make that out?'

'Well sorr, if there was any work attached to it you'd be sending for me.'

* * * * * * * * *

A date with a tigress

Lt.Colonel Wilson's CO of the 19th Lancers told him this tale of big game hunting in India which is worthy of repetition here. The hero is an elderly Colonel – once a great shikari and not so unsuccessful with the ladies either – but now past his prime he was rather losing his grip and apt to lose the thread of a conversation.

However, seated on the mess verandah under the only punkah he was regaling some young officers with a description of his first tiger shoot. They sat round fascinated as he described how he was sitting in a machan built in the lower branches of a tree overlooking the corpse of a water buffalo that a tigress had killed the night before. He was expecting her back to eat the remains – at any minute.

He heard the sound of twigs and leaves crackling and slipping his safety catch forward he peered into the moonlight. He saw nothing; but suddenly he heard the sound of a branch breaking behind him. The tigress had climbed into his tree and was right behind him!

'I felt her hot breath on my neck', he whispered dramatically. The young officers sat agog, but at this critical moment a khitmagar entered and announced 'Telephone Colonel Sahib bahadur'.

With jingling spurs the old Colonel shambled off to the mess office. The caller was an old flame of his younger days. It was a long conversation and when at last he rejoined the company every thought had left his head except his recent telephone conversation.

'What happened next sir?' prompted one of the officers.

'Eh what? Where was I?'

'You felt her hot breath on your neck sir. What happened then?'

'Well, of course, m'boy I went to bed with her.'

* * * * * * * * *

There is justification for recalling the story of a new year's eve guest night on the Frontier in 1905 told by F. Yeats Brown in his book, Bengal Lancers. Well-trained servants, he said, appeared in the ante-room as if by magic to remove all breakable furniture and replace it with a special set of chairs and tables made to smash.[15] A fat Gunner subaltern was swaddled in an enormous roll

15 The officers' mess at the Cavalry School at Saugor, Central Provinces, used to be the scene of wild guest nights.

If an officer smashed an arm chair or table, he had to supply an unbreakable one in it's place, and each piece of furniture thus replaced had a small silver plate affixed; for example 'Presented by Captain J.Smith of the 37th Lancers (Baluch Horse) 1st January, 1912'.

Lt.-Col. F. Wilson commanded the Indian wing of this establishment when it became an Infantry School during the last war and the furniture was still there.

of webbing, there were wrestling matches, a boy in the Punjab Frontier Force brought in a little bazaar-pony and made it jump sofas. He had his trousers torn off. Yeats Brown recalls standing on his head to prove that it was possible to absorb liquid in that position, and the festivities included billiards fives.

There is also in Yeats Brown's book the story of the elderly major who had celebrated his impending departure with more than enough champagne and laid on his bed in something of a stupor. By the flicker of a nightlight which always burned beside him he began to conjure up visions and saw, at the foot of his bed a skinny outline that moved every time he did. His trembling hand seized a revolver and he fired – immediately roaring with pain, for he had shot his own foot!

* * * * * * * * *

The Dreaded Parallels

It was from Major Tommy Gilfillan of the Cameronians that Littlewart learnt how his battalion had suffered from the dreaded parallels whilst serving on the Persian Gulf in a Sheikdom which consisted of limitless acres of sand covering countless gallons of oil and driven over by miles of Cadillacs. They were in a tented camp and by any standards it was a rough station.

To prevent their pipes becoming furred up with sand it was necessary for the men to lubricate their throats from time to time with remedial medicines most to the taste of Scotsmen. There was one very large officer whose throat was always sand paper dry – a hard case – who used, however, to take his medicine like a man – in the officers' mess and out of bond. At times he might take more than the prescribed dose and find himself unable to make his way back to his tent. Word would then be sent to his batman and in the distance in rich Scots the cry would go up 'It's the dreaded parallels!'. The call would be answered by his 'mucker' and together they would come to the officers' mess and haul the inanimate bulk to his tent – one to each arm like a couple of tugs salvaging an ocean liner. His legs would, of course, drag behind in the sand leaving the unmistakable trail of the dreaded parallels.

The disease was not confined to the officers and the MO on morning sick parade might find himself diagnosing a mild case of DPs and being a Scot himself and an understanding man giving the sufferer one day's 'bed down'.

The only sure cure was a UK posting, preferably before a severe case of DPs developed into DTs.

* * * * * * * * *

Deathly hush at breakfast[16]

Clanger was not the only 2/Lt. whose attempts at hearty conversation with senior officers at breakfast met with the retort not too courteous. However, nothing could have been more considerate than the handling received by one of Clanger's friends. The following conversation took place:

Gushing 2/Lt: 'Good morning sir.'

Crusty Major: (grunts).

Gushing 2/Lt. bent on a suicide course: 'Hunting today sir?'

Crusty Major (calling): 'Waiter.'

Waiter (respectful but with a good idea what is to come): 'Sir?'

Crusty Major: 'Mr.Butcher wants to talk; talk to him.'

To which, had the incident taken place in a Guards officers' mess, the reply would have inevitably been, 'SIR'.

16 Here again Littlewart has assumed the identity of Lt. Colonel Wilson.

A 'Shooter Sowar' takes out beer and 'sandbeefs' to members of an Indian Cavalry Mess on a duck shoot.

See page 117.

Deathly hush at breakfast[17]

However not all such attempts at conversation were diverted to the mess waiter and in one infantry regiment the senior dining member felt that the mess as a whole should be treated to the conversational brilliance of the silence breaker. In this regiment it was the custom not to speak to a young officer for several months after joining. One young officer, endured this ridiculous state of affairs without complaint until one night he found himself, to his horror, seated at dinner next to the senior dining member – an elderly, crusty, bachelor Major of the type that was not uncommon in those days.

Solemnly the subaltern munched his way through four courses, and watched the port go round without either addressing a word to his neighbour or being addressed in turn. Finally he decided it would be uncivil in one so young not to say something to his very senior dinner companion. So screwing up his courage he turned to his neighbour and said, 'It has been a lovely day hasn't it?'.

The senior dining member turned slowly in his direction and stared incredulously without saying a word. Then picking up his dessert fork, he hammered the table with its handle. Conversation stopped and all present looked at the senior dining member.

'Silence please, gentlemen,' he said, 'our latest joined subaltern wishes to address the mess!'

Poor lad, he left the army shortly afterwards and who would blame him.

Deathly hush at breakfast[18]

There is, of course, another story about talking at breakfast which was that when the newly arrived subaltern came in he foolishly said 'Good morning' to the Senior Major, who replied: 'Good morning, good morning, good morning, good morning, good morning, good morning, good morning – and let that last for the whole bloody week.' There is an alternative to that story in that the Major replied: 'The morning's all right, it's the people you meet.'

* * * * * * * * *

Conversation in the mess at breakfast is often restricted to bare essentials such as 'Pass the salt old boy' or 'Give the sugar a fair wind Clanger', but when conversation does break out it is usually of a cheerful nature. A remark once made by the mess sergeant of the 1st Bn The Essex Regiment, however, effectually put a damper on all further cheerful conversation. One morning he entered the dining room, approached the adjutant, who was a living-in member, bowed slightly and with no more emotion than if he was announcing that he regretted that the milk had gone off informed the adjutant that 'Beggin' your pardon please sir, but there's a man hanged himself in the lavatory'.

And on that note it might not be inappropriate to sound the Last Post and close this chapter.

17 The tale is told by Major-General J.D. Lunt, CBE.

18 From Lieut.-General Sir William Oliver GBE, KCB, KCMG, DL.

ix

Some not too beastly customs

Brigadier Littlewart had had a tiring day visiting a static unit of his command in UK and he now entered the ante-room of the officers' mess (he was staying the night before going on to another unit in the morning) badly in need of a whisky and soda – long and strong. He pressed the tit and sank thankfully into a deep armchair narrowly missing the mess cat. Thank God they don't allow dogs[1] in here, he thought, or I might have been savaged by a doberman. Minutes passed and nothing happened, so groaning he heaved himself out of his chair and rang again somewhat forcefully. After a further interval a soldier – he took him to be a soldier though neither his hair style nor footwear gave a clue to his calling – appeared buttoning up his white jacket.

'Did you ring sir?'

'I did indeed. I'd like a large whisky and soda'.

'Sorry sir can't do it.'

'Why not?'

'Roddie – er beg pardon sir – Mr.Jones hasn't arrived yet.'

'And who is Mr. Jones?'

'The civilian barman sir. He doesn't open the bar till 6.30.'

The brigadier glanced at his watch. God! a quarter of an hour to go. He wished he had not insisted on staying in the mess 'to test it' instead of staying with the CO. How he missed the days when you pressed the tit and a soldier brought you what you wanted. It still happened in some messes – but too few.

'Is it impossible to get anything now?'

''Fraid so sir, y'see he's got the keys. I've a Coke in my bunk, but I don't suppose you'd care for that.'

'You are quite right, but it is kind of you to offer. I'll just have to wait.'

The man left and the Brigadier huddled deeper in his chair muttering imprecations and wishing he was at his own HQ where he tried to maintain a military flavour in the mess. It entailed a perpetual battle against establishment committee proposals for the inevitable cuts in military manpower and suggestions for more civilianisation. He realised it was a forlorn rearguard action he was fighting but he was going to fight every inch of the way.

1. Many messes did not permit animals but dogs were often allowed at the discretion of the PMC and if they were well behaved. So reads the rule applied in the King's Regiment and probably many others. It would seem from the photograph of a group of mess members of the Queen's Own Hussars that even officers were allowed in the mess if they were well behaved and sat on the floor!

Even officers were allowed in the mess
if they were well behaved and sat on the floor!
Major D.C. Risby – Queen's Own Hussars

He sighed. Can't turn the clock back, but he could not help thinking nostalgically of those happy days in the Ramrods' mess, between the wars, when the army preserved its own way of life and did not want to be equated with civilians – damned box-wallahs! In his reverie he dreamed of a typical day as a subaltern in his regiment, and many another like it, in those long gone days.

* * * * * * * * *

He would be called by his batman at around seven o'clock unless there was an early parade – the exception rather than the rule – and it would be his own batman too, no sharing in those days or cleaning your own kit. Without 'stay bright' buttons and plastic or patent leather an officer's uniforms and leather needed all the attention a personal batman could give, apart from cleaning his quarters, often a little sitting room and separate bedroom.

Duly accoutred in service dress – those comic plus fours and puttees! – he would go to the mess for breakfast. This was a substantial meal with the choice of three or four dishes sizzling on a hot plate (which actually worked) in addition to porridge or cereal; and there was no objection to sampling more than one dish. The meal would not be eaten in a deathly hush, but on the other hand it was not notable for gay chatter. One reason being that many an officer would be trying to read a newspaper propped on a stand in front of his plate whilst at the same time doing justice to his breakfast; a task quite sufficient to occupy all his attention without trying to indulge in small talk as well. (Some regiments read papers at breakfast, in others it was forbidden.)

Littlewart remembered ruefully the day when he had the madness to seat himself in the empty chair next to the 'living in' major, an extremely crusty bachelor just about to be retired on age and with a large chip on his shoulder for being passed over for command. Not a type to be dealt with lightly at the breakfast table, but fools step in..... Littlewart had even gone further in his folly and had cheerily greeted the Major with.....'Good morning sir, lovely day, Spring in the air.'

To which, to the great delight of the junior officers, he received the withering broadside in reply.....'Why the hell should I you bloody fool?'.

Littlewart learned that, even if there are no other empty places available, it was preferable to wait for one than encroach on the Major's purdah.[2]

Breakfast over, the day's work would begin with company office when the company commander – often Littlewart himself – would deal summarily with offenders or remand them for the commanding officer as appropriate. Then would follow inspection of barrack rooms and it would be a surprise if his platoon mustered more than a dozen soldiers. After that, probably weapon training and those twelve loyal heroes – possibly depleted by one or two 'detailed for duties' – would practise stoppages on the Lewis gun for the hundredth time. How did it go? The Brigadier taxed his memory. Yes, he thought he had got it..'Gun's firing all right – gun stops second, won't fire third. Then 'Mag off – UP; clearing plug number 2; legs number 2; mag ON gun's firing all right'. Triumph! This stoppage was invariably presaged by the caution from the instructor of ''ead away take no notice what I'm doing now' and of course the trainee knew at once what he was doing and at once guessed the stoppage. He was inserting the 'separated case'. How bored Littlewart was; it had not taken him long to learn how to strip the Lewis gun blindfold in about thirty seconds! So, exhausted by about an hour of this he might stroll down to the mess for a bracer; a sherry and biscuit or even two or possibly a beer or even coffee, but not TEA. Tea! Tea was drunk in the company office with the CSM and CQMS. It could be almost as much as his commission was worth to ask for it in the mess for elevenses. It might be suggested that if he wanted tea he'd better transfer to another regiment or to one of the services, but tea in the Ramrods!..... In those days the complete shut down of a unit for NAAFI tea break did not occur.

The second half of the morning would follow a pattern similar to the first but the tedium of the routine might be enlivened on Fridays by a trip to the bank to draw the company pay.[3] There one might meet other young officers similarly engaged and possibly repair to the officers' club for a quick one before returning to barracks to spin out the last half hour or so before foregathering in the ante-room for lunch. Not an arduous morning, but then what could you do with your brave band of twelve that had not been done countless times before? Real soldiering took place in the foreign battalion which would be, if anything, *over* strength.

Whilst it was not obligatory to go in to the dining room more or less together it was not unusual, if for no better reason than not to let the meal spoil and to try to get a bit of the 'undercut'. The standard of messing in the officers' mess of those days was as high as it was low in the men's dining hall. The creation of the Army Catering Corps has been the only revolutionary change in the army for centuries! Only officers detained on some duty, such as orderly officer, would dare to incur the mess sergeant's wrath by coming in late to lunch. There were no prolonged lunchtime gin sessions, even on Saturdays. No doubt the 'no treating' rule which was pretty religiously observed in most messes contributed to this, coupled with the fact that the CO was liable to scrutinise mess bills to check that young officers were not spending too much on wine. And of course, in the thirties bars were anathema; forbidden by *King's Regulations* in fact, whereas they are now an inescapable feature of mess design.

At lunch there would be a choice of joints over which the mess sergeant would preside and discourage young officers from trying to carve until they had been passed fit by him. Littlewart blushed at the memory of a joint flying through the air to the carpet as a result of his efforts to get at the undercut!

The afternoon lived up to the popular saying of 'half a day's work for half a day's pay'. After all, ten bob a day even at the value of those times was scarcely a princely wage, especially if one took into account the principle, which still exists, that if an officer is not on leave he is on duty – for the full twenty-four hours of the day. If the officer was not engaged in sport of some kind,

2. In perhaps most messes there were no chairs in the ante-room which were tacitly reserved. But in the Devonshire Regiment there was the field officers' fire-place and the subalterns' fire-place and both were sacroscant. Presumably captains were not left out in the cold and joined the senior fire-place!

3. There were in those days no unit paymasters and each company ran its own pay account maintained by the 2IC and the CQMS.

A Station Mess in India at the end of the last war, where British and Indian Officers are enjoying their drinks at one of the 'new fangled' bars which were being introduced at that time.

See page 119.

which did fill many an afternoon and especially Saturdays when the big matches were played, he might take the opportunity to pay a few duty calls and drop cards on the marrieds.[4] *Engraved* cards, of course! Printed cards, like made up evening bow ties, were simply out of the question; only bounders like Flashman[5] would dream of using either. Brigadier Littlewart wondered how many plates he had had made in his time to be discarded on promotion or lost in moving stations, or for some such reason.

When he first joined, tea had been served in the ante-room, an individual tray for each officer with its own tea pot, milk jug, sugar, four slices of buttered toast or bread and butter and a slice of cake. It was quite a while before 'schoolroom tea' in the dining room was introduced. Of course, even if you had been playing games you had to be properly dressed; a blazer and scarf were the usual rule, even a high necked sweater without a jacket was not regarded as quite the thing, and 'tinker's mufti', a combination of uniform shirt and tie with grey flannel bags and a sports jacket, was unthinkable!

Between tea and dinner he might play a game of billiards or find himself armed with mess cards and accompanied by another officer detailed to pay an official call on another officers' mess, (preferably the Royal Scots Fusiliers – they don't deal in chota pegs, only burra!). They would arrive thirstily around six o'clock in the hope of finding someone in, if only the orderly officer holding the fort in the hope of welcoming such an invasion into his dreary day.

Alternatively he might go 'poodle-faking'[6] or even work for his promotion exam. A lot must have done so as, albeit with the aid of the Metropolitan Services College, most of them seemed to pass first go. You had to, you could not afford a second course!

Then at around 7.30 the ante-room would be aired; windows opened to let out tobacco smoke (no more smoking would be allowed before dinner. This dates from the days when the dining room and ante-room were one big room divided only by a curtain); and it was expected that no one would be found there any longer in plain clothes. If he were, a playful effort might be made to eject him forcibly as..... The Brigadier awoke with a start. The mess servant was gently shaking him.

'It's seven o'clock sir. Mr. Jones has opened the bar. Like me to show you where it is. You'll have half an hour before supper starts, sir'. Before supper starts, thought Littlewart; supper at seven thirty. Good God! But first things first and the first thing was a great big whisky and soda. If personal service in the ante-room was no longer possible and the fountain would not come to Mohammed, Mohammed would have to go to the fountain.

'Thank you' said Clanger. 'Lead me to Mr. Jones' oasis in this dry pseudo-military desert.'

4. The newly-joined officer of whatever rank had to call on the married officers and their wives, which meant putting on a presentable suit and ringing the front doorbell between three and four of a week-day afternoon. It was hoped that the recipient of the call would be out and a good deal of contrivance took place on both sides to bring about this desirable state of affairs, for the afternoon call was just as much an ordeal for the young bride as it was for the bashful bachelor.

   They ordered these things better in the East, where each bungalow had a box at the gate where visiting cards could be left. Even this had its dangers, recalls Major General Goldsmith, as when one young officer entrusted with the cards of his friends as well as his own, instead of placing a proper selection in each box, deposited *all* his own in the first, *all* those of his friend A in the next and so on. And then again there was the young officer who left twenty of his own cards on the Brigadier and mem-sahiba and was asked to dinner ten weeks running.

   This sounds as if the young officer could only have been Littlewart, but as we have seen, he was getting on by the time he went to India.

5. Young Harry Flashman, grandson of the notorious Victorian VC general, had been at Sandhurst with Littlewart. He had joined a regiment that scarcely deigned to acknowledge the existence of any regiment below them in the Army list. He had looked forward perversely to being the only bounder in that exclusive lot but he wrote to Littlewart complaining that 'dammit the regiment was full of them and he was considering transferring to his Grandfather's old regiment' where he felt that he would surely achieve his aim. But would *they* have taken *him*? Doubtful!

6. The origin of the term 'poodle-faking' is wrapt in mystery. It merely meant taking a girl out, not necessarily courting or with the intention of making a dishonest woman of her; just for pleasant company. The other ranks, whatever their intentions and possibly they were not always on so pure a level, called it 'square bashing', an equally inexplicable term.

The waiter had no idea what the Brigadier was talking about — quite a card he reckoned him — but he led the way to the bar and the Brigadier was happy to follow. There let us leave him restoring his tissues with the able assistance of Mr. Rodney Jones whilst we take advantage of the interval before the evening meal to study some of the more unusual customs observed in some messes.

* * * * * * * * *

In this category, first mention must surely be made of the Customs Book maintained by the Gloucestershire Regiment. Though possibly not a TOP SECRET document it is at least FOR REGIMENTAL EYES ONLY! In it are recorded customs dating back to the founding of the regiment in 1694. There are also very modern ones since from time to time are added newly-born customs which it is intended to keep alive. Needless to say the procedure for proposing the inclusion of a new custom is surrounded by much regimental red tape.

The Keeper of the Book is the 2IC and only he, or in his absence the president, may send for it, for it is always kept close at hand at guest nights which are a frequent source of the creation of new customs in any mess. The Peninsular War provided much material in addition to the particular manner of proposing the loyal toast and the habit of bivouacking under the dining table, customs which are both mentioned elsewhere in this book. Many regiments keep scrap books which make interesting unofficial history. The subalterns of the 2nd Seaforths kept a 'raspberry book' and the names of officers who had earned this distinction were recorded in it. They include a Field-Marshal.

The 1st Essex kept a similar book. In this battalion a raspberry was called a 'ptarmigan' — don't ask me why — and the book was called the Ptarmigan Ptimes. In it were recorded all sorts of regimental gossip, rumours and scurrilous incidents, but regretably no Field-Marshal's name.

In some regiments special customs have been established for celebrating promotion to captain. Not without good reason since prior to the Hore-Belisha reforms just before the war it took as long for an officer to obtain his captaincy as it took a soldier to earn the 'Rootie gong' which was awarded for 18 years service of undetected crime. The 1st Battalion The Loyal Regiment possessed a beautiful silver chalice known as the 'Subaltern Cup'. There are no indications on it of its origin but there is little doubt that it was 'acquired' during the Peninsular War. On promotion to captain the ex-senior subaltern hands over the cup to his successor after dinner in the mess when all the senior officers have left the room. The cup, which holds nearly half a bottle, is filled with champagne and the newly promoted captain has to drain it at one draught. His successor as senior subaltern, followed by all the subalterns in turn, then do so at the captain's expense. Should any officer fail to drain the cup at one draught the procedure is repeated all round the table at that subaltern's expense. This may sound like a Babylonian orgy but remember that it would happen very seldom in an officer's service. Possibly, owing to cross postings, secondment, service at the Depot and so-on, only once. Promotion came almost literally as slowly as a funeral hearse since often death in the senior ranks was the only way of creating a vacancy and subsequent promotion to fill it. Nowadays with timed promotion the excitement of making captain is small; if the regiment is under strength it may not even excuse the new captain from doing orderly officer but in the thirties it really was an event to be celebrated even if the first month's additional pay went in the celebration! To the man who could serve efficiently and, taken all round, fairly cheerfully I believe there were fewer chips on shoulders pre-40 than post-50 for seventeen years or so with no increase of pay, head-dresses should be removed and three hearty cheers given.

This leads logically to the custom of wearing hats and belts in mess. A guest to lunch at Wellington Barracks may notice that some officers may be wearing hats or belts or both. This does not denote that they are piquet officer, field officer of the week or anything of that sort. It is a custom in the Guards battalions that served in the Crimea that they may wear hats and belts in the mess up to six o'clock, and originates from that time when except for very senior war lords living aboard ship in the harbour, accommodation was limited to tentage. And tentage is notoriously lacking in any facilities for hanging up coats and hats. Officers therefore kept them on. Nowadays the custom is entirely optional and the officers seen wearing hats and belts may only be doing so because there is no room left on the pegs in the hall which have all been taken up by previous arrivals — or guests!

Another suggested origin of the custom is that during the last century gentlemen wore their top hats in such places as the House of Commons and their London Clubs. It therefore became a

habit to wear them in this sort of institution and the custom was carried into the officers' mess. Either explanation is equally credible. I would put my money on a compromise. I picture a Guards officer, returned from the Crimea entering the mess in uniform. He sees there already hogging the best armchair a Whitehall warrior who had dodged the column – wearing a topper. Incensed, 'the hero of Inkerman' goes back into the hall and puts on his uniform cap and returns silently but obviously saying 'So there, sir'. The retort courteous, silent but obvious. Somewhere in between the explanation must lie.

A similar custom was observed by the Highland Light Infantry and the Royal Scots Fusiliers now amalgamated as the Royal Highland Fusiliers. An officer might wear a glengarry in the ante-room during the morning and a dirk may be worn in the mess but Sam Brownes and claymores must be removed and left in the hall. The origin appears to be forgotten but possibly retention of the dirk might have provided the wearer with some form of arms. This certainly is the origin of the custom in the Worcestershire and Sherwood Foresters Regiment, whereby the captain of the week and subaltern of the day wear swords at dinner. One night in September 1746 the officers of the Worcestershire Regiment were at mess in their station in North America when they were treacherously attacked by Red Indians who were supposed to be loyal. The attack was beaten off but to guard against similar hostilities in future the custom of wearing swords at mess was instituted. This continued as a regimental custom after the Regiment had left America, but in about 1850 a colonel who had little regard for these things abolished it, with the concession that the captain of the week and the subaltern of the day continued to wear their swords at mess. This custom was maintained by all battalions in peace time. The Worcestershire Regiment was the only regiment in which this custom had official sanction and it is maintained today in the Worcestershire and Sherwood Foresters Regiment.[7]

7. In many messes, however, to wear a sword in mess was not prohibited and many were the gambits employed to snare a young officer into drawing his sword in the ante-room, the penalty for which was drinks all round. It was suggested to Littlewart that possibly the blade of his sword was not properly engraved in the manner peculiar to the Ramrods and that if it was not he should send it to Wilkinsons to have it done. He fell for it like a lamb and it was 'Press the tit Tom, it's drinks on Littlewart'.

Some regiments were stricter than others. A contributor writes: 'It was not until the outbreak of the 1939-45 war that I learned that one was not allowed to wear a sword in the ante-room. This rule was apparently introduced by the very old reservists who were sticklers for etiquette. Not that we had any swords to wear (i.e. in war) and there was a call for drinks all round when I came into the ante-room with a belt round my waist. In peace time I had been in the habit of taking my morning sherry and digestive biscuit with my sword still fastened to me and nobody objected'.

A Christmas dinner on the Heights before Sebastopol - Crimean War.
3rd Bn. Grenadier Guards 1855.

There were also customs from which the other ranks contrived to benefit. The officers' mess was the home of the colours and following a parade on which the colours had been carried, after they had been cased and 'laid up' in the dining room, the colour party would be entertained to drinks in the back premises of the mess although their duty had been no more arduous than that of any other soldier on the parade. The chaps who really carried the can were the officers who carried the colours. It was no joke in a bit of a wind and especially if the colours were the old pattern, large-sized variety. The 1st Bn Essex Regiment had these and they were reputed to be the last carried into action; at the battle of Taku Forts in China. It was the custom for the officers to present the drum-major with £1 for the privilege of having carried out the duty. The drum-major was the 'custodian' of the colours; he supervised their uncasing in the mess prior to the parade and their return and casing after it. There does not appear to be any particular reason why the drum-major except that he was, as it were, a spare number. In war when the bandsmen were made stretcher-bearers he became the intelligence sergeant, and six of the drummers, who were almost without exception the stupidest soldiers in a battalion, became the intelligence section. How Littlewart suffered when he became Bn Intelligence Officer not only from his section but from being mounted on a GS bicycle, quite the solidest and heaviest piece of equipment in proportion to size that the army kept! When the bn recce party set out on a task the CO, adjutant and signals officer all had chargers, but the Bn IO had to do his best to keep up on his flat feet or iron horse – with a flat tyre. Incredible though it may sound this was the situation pre-1939. At least in those days the bandmaster did not become second quartermaster. Now, of course, things are very changed and there is in addition an Intelligence Corps, which does not recruit from out of work jazz drummers!

The brigadier was called in the morning by a civilian of military bearing, who might have been the twin of the 'general' who had looked after him at Sandhurst so many years ago, and whilst he sipped his tea he ruminated on the batmen it had been his luck, or more often bad luck, to have. Why batmen? he wondered. They did not bat nor did they bowl and they were often not much good in the field either. He had suffered from many bad batmen just as Lt. Lukash of the 91st Austrian Infantry had suffered in the 1914/18 war. It had always been his luck to get hold of the most objectionable batmen. They came and went continuously and at last he would sigh to himself when a new one arrived.

'Here's another low brute palmed off on me.' In fact what he got was the good soldier Schweik[9] whom he won from Padre Lucina playing pontoon! It was not so very different in the British army. There were few batmen of vocation and those that there were the adjutant and the married officers managed to acquire. Naturally, not the best soldiers were detailed as batmen. They were too useful in the company to be wasted on a single officer.

There had, he remembered, been that prize idiot Pte. 'Froggy' French. At the time Littlewart had a dog – or rather puppy – and it had made a puddle on the floor. So he told French to get some pepper and put it down. When he returned to his room for lunch he found that Froggy had got round the cook sergeant and had spread a thin layer of pepper over the entire floor.

Then there had been Sheldrake, the armchair athlete. Clanger himself had been a keen athlete and would be out training on the track every day. Once on his return from leave he learnt that Sheldrake, who normally did not stir a finger or toe, had been seen running round the track daily – doing Littlewart's training for him! Not only that, he had as loyally borrowed Clanger's running kit to do it in!

Ellery the Feet had been far the worst. He had a squint and a touch of the tarbrush. These could be ignored, but he also had feet which could not, within four walls and even at close range out of doors. Fortunately he got himself into trouble and was returned to duty when Clanger was despairing of getting rid of him. He had 'borrowed' Littlewart's car when he was on leave and had contrived with a pal to have a minor accident – fortunately without damage, but it led to an amusing court case. Littlewart was required to attend and duly reported to the local court in service dress, Sam Browne and sword. (He would have worn medals if he had any.)

8. The servants at the RMC and later RMAS were always called 'General'. Many looked the part too, though of rather Boer War vintage!

9. Students of army life and particularly the genus 'batman' are recommended to read that delightful satire of the Austrian army by Jaroslav Hasek called *The Good Soldier Schweik*.

1940
The Mess at Armentieres . . .

. . . the little café opposite.

As chief witness for the prosecution, he gave his evidence; that he had never given permission for the said car to be borrowed by the accused whom he now recognised; that said car was insured only for himself driving and that even if it had not been he would not have loaned it to the accused. In fact he gave them pretty black characters but no less than they deserved. The bench thanked him for his clear and valuable evidence and stood him down and went into a huddle rather like a scrum in American grid-iron football, and to complete the simile, from time to time snatches of phrases could be heard such as ..... '.....evidence of character.....', 'prisoner's friend.....', 'quite young fellows.....will the army take them back.....', and so on. Eventually they broke up and the chairman enquired whether there was anyone in court who could speak for the accused's military characters. 'Yes', said Littlewart rising, 'I am here to speak for the accused.' (Sensation in court.) At the time his company was detached on musketry camp and another officer simply could not be spared. So back he went into the witness box and gave the accused excellent[10] military characters, pleaded that it was their first offence, said that the army was willing to take them back and, eyeing the chairman of the bench meaningly, that their accounts were more than two pounds in credit. Once more he stood down and the bench went into a huddle again – '.....thirty shillings.....but no licence or insurance.....three pounds.....quite young fellows y'know.....forty shillings.....'. At length they settled for forty shillings and the accused were duly convicted. But the fun had by no means ended. The money had to be paid into court and the accused pleaded that they had not got the needful. When the clerk was getting a little nettled Littlewart once more stepped into the breach and said that he was prepared to pay the fine. (Prolonged sensation in court and some laughter.) This he did, took his receipt, left the court and drove the accused back to camp in the said car, thinking that the court would never be the same again. But it was. A little while later he was pinched for speeding in a built-up area, appeared before the same court and was fined.....forty shillings! But it had all been well worth it to get rid of the noisesome Ellery.

It took the 1939/45 war to bring Littlewart the best batman he had ever had. On mobilisation he had been appointed to a newly-formed HQ and had to select a batman from a line of reservists drawn up like an identification parade. He had chosen Pte.Anscombe. What an inspired choice. Anscombe turned out to be a veritable Jeeves; in fact he had in civil life been a gentleman's gentleman. He had a great sense of humour. In the retreat to Dunkirk the mess had been established for a short while at Armentieres and Clanger, who had been on the advance party, discovered that the cellar door had been left unpadlocked. As wines member he at once gave orders that the contents were to be issued only on his orders. So at dinner, in order to wash down the corned beef rissoles, he said to his batman, 'Anscombe, I wonder if you could pop across the road to the little café opposite and get a bottle of red wine'.

Off Anscombe popped and after a suitable interval would return with a bottle from the cellar, and perhaps one for a friend of Clanger's too. This continued happily for a very short while until the time when, to add a little Gilbertian versimilitude, Littlewart gave Anscombe a note with which to pay. When he returned Clanger asked for his change. Without a smile or batting an eye Anscombe replied in the best Jeeves manner. 'I regret to say there is no change sir. Whilst I was there I took the liberty of buying two bottles.'

There was nothing Littlewart could say and Anscombe got away with a pocketful of small change! Unfortunately on return to England Littlewart was posted and Anscombe remained to become mess corporal for the rest of the war doing the job he knew best to everyone's satisfaction. What more can a man do?

At the moment the brigadier had a good batman but as he was to be away for only one night and there was very little kit to look after, his wife had not allowed the even tenor of the household to be disturbed by him taking away *her* batman even for one night. The brigadier did not complain. His staff brought him enough troubles without adding to them with strife on the distaff side!

10. The necessary qualifications being what they were, it was often very difficult to give a soldier an official military character other than 'good'.

X

After dinner diversions

'Tis noise and nonsense are their dear delight
And stupid pleasures crown the drunken night.'

These lines were penned in by an anonymous contemporary poet to describe the diversions of Bath in the eighteenth century. They might well be applied to many a guest night that got a little out of control – they sometimes do! But a successful guest night can be a quiet affair; most weekly 'band' nights were, and all 'ordinary' nights. Of course, there were now and then those nights to which the couplet could quite fairly be applied.

Both the 'Shop' and Sandhurst gave a taste of what could come – harmless diversions such as larding the tram lines outside the barracks at Woolwich to the great discomfiture of the trams and little less to the occupants; or the wild 'cyclocade' as it might be termed in modern jargon, of cadets returning to the RMC after attending the Arcade cinema at Camberley. Cyclocade is scarcely the word though; there was no PRO to decide the order of your going according to protocol, it was just a wild gold rush, as about a hundred Gentlemen Cadets pedalled into the staff College and RMC grounds – in by the OUT gate naturally – and pounded flat out towards the college to get there before the dining out lists were closed and taken away. Some with lamps, some without, some with torches and some with what in *Small Arms Training Vol* whatever it was that dealt with the revolver, 'the instinctive pointing sense'. The leading file had to keep their eyes skinned for the WD policemen alert to catch a GC without a light and if one was spotted they raised the cry 'Bluebottle' which was passed down the column so that the GC still searching for his bicycle outside the cinema was duly warned. The gentlemen cadets would already have made the presentation of the film a bit of a chaos by the rather ungentlemanly behaviour of giving a running commentary, sometimes lewd and ribald but often very witty. It might be a Greta Garbo/ John Gilbert love affair and a passionate scene between the two was really only improved by the comments of the cadets. They were, of course, the days when films were otherwise mostly silent. Though some complained, I believe that many of the citizens of Camberley went to the pictures to enjoy the GCs' comments rather than the film.

Or perhaps if you were in 'Lovely One' you might suddenly have the uncontrollable urge to throw a bicycle rack through the ante-room windows of '..... Four'. You were prepared to pay afterwards, of course, and did willingly, (although you had to pay to go to Sandhurst at all). So often today it seems that 'rag' participants, in all walks of life, expect to do considerable structural damage and not to be called upon to make it good. A grant should be made in aid.....

An officer of the Devonshire Regiment recalls of those occasions, 'I have known a serviceable settee (officers for the use of) to be reduced to a pile of wood coiled springs and stuffing in twenty five seconds dead. I remember the occasion well because a furniture nail went through the sole of my mess wellington into my foot. I have seen my divisional commander divest himself of his bemedalled coat and, flinging it into a corner, dive into the fray yelling like a banshee. He insisted on paying his share of the cost of the furniture and fittings which had been destroyed'. Quite rightly too. You take your fun where you find it and should be prepared to pay.

On an ordinary dinner night, however, the ante-room would be a quiet place. The attendance at the meal would not be at full strength allowing for young officers dining with marrieds, going to the pictures, 'poodle-faking' and the like. The officers who had dined might play bridge, billiards or just read the papers or gossip over a slow drink. There was always the possibility – dread thought – of having to attend an all ranks' dance or a sergeants' mess function of some sort or other. The officers would arrive – pretty well primed – after dinner and would be expected to leave around midnight. Of course there would be little or no dancing. You had naturally to dance with the wives of the RSM, your CSM and CQMS, your platoon sergeant, maybe the company storeman as a gesture to probably the oldest soldier in the hall (gymnasium), and the WT sergeant's wife if you were Weapon Training Officer and so on.

There was a certain protocol among the wives as a subaltern of the Royal Scots Fusiliers discovered when he asked his CSM's wife for a dance. 'Thank you no, Mr.Buchanan,' she replied. 'I am keeping myself cool for the company commander.' Which might have been interpreted by Mr. Buchanan as a backhanded compliment!

With these few dances done, the terpsichorean duties were over and there followed the harder task of propping the bar and 'chewing the fat'. Often most pleasant, but at an all ranks liable to be too tedious if caught by the company bore and forced to listen to dreary reminiscences or embarrassing eulogies. – 'You're the best officer we've ever 'ad in 7 Platoon sir.....' – 'We should 'ave won the company fubball cup sir.....that wasn't an offside goal.....you're the best forward in the battalion straight sir 'strue.....' – and so on and so on until why, hullo the CO has gone, time to be moving. Unlike the modern day when the young bachelor officer often brings his own 'bird' or if his cage is empty at the time tries to cut out his platoon sergeant with that saucy piece from the supermarket. And stays till dawn! Bar propping with the company bore is a preferable form of democracy.

Or it might be a sergeants' mess games night. Here teams would be formed for darts, shove ha'penny, snooker, cards or just straightforward drinking! The games were played on the American Tournament system and as the night drew on the officer who had opted for the drinking team might have regrets as the last black stayed up for twenty minutes.

These nights were much more dangerous than an all ranks' dance naturally: it would have been a poor show if they had not been. The general aim was that all games should end in a draw and that at the end of the night all concerned would be equally anaesthetised. And the mornings after were much greyer when death's sting would have been very welcome – at least it would be quicker. It was a long time to eleven o'clock! The RSM, CSM, platoon sergeant, or whoever it might be, would take particular delight in reporting the parade present in a voice like the crack of doom scented with the concentrated blast of a hundred breweries! At least the officer could take consolation from the thought that 'it hurts him more than it hurts me!'.

On the weekly 'band' night the tempo might be stepped up a little. The attendance by unmarried officers would be 100 per cent, a married officer or two might dine in and there might be some private guests. Some of the wilder games usually reserved for regimental guest nights might even be played but it was on the regimental guest nights that the hair was really let down and the wily subalterns took evasive steps to avoid being co-opted into bridge fours with senior officers. The diversion then did not take any set form. As a contributor from the 'Skins' so happily puts it. 'The behaviour on such occasions is of the as-required variety using such props as gold-fish ponds, chandeliers and minstrel galleries as the spirit moves us.' It could not be more aptly summed up. But whatever happens, it is expected that the subalterns should be the instigators and leading lights. So fall in the subalterns and let's hear from you.

As we have already learnt, in the Devonshire Regiment a toast was drunk to the French army. At one time there was an adjutant who was not particularly Francophile but happened to be the proud possessor of a Rumanian decoration, and at times the inspiration would move him to stand on the fender in the ante-room and take the salute as the subalterns marched past. This would only be the prelude to such activities as – and this is recalled by Brigadier Leeper – throwing Bernard Fergusson out through the window. He reports that he bears the marks to this day but does not give the reason for this drastic treatment. But his brief flight through space may have helped him to his meteoric career which has probably brought him as many Gs as there are Ss to his name.

Tables prepared for a Regimental Guest Night – RA Mess, Woolwich. Mess servants in powdered wigs – 1973.

At Catterick it was a popular pastime of the junior officers of the Royal Tank Corps Battalion to dive through the ante-room window – open of course – into the flower bed below. No harm was taken until one of the star divers did the same at a guest night in a strange mess and landed after a six foot drop on a tarmac car-park. However he fell relaxed – he had been 'out of pain for some hours' – and broke no bones.

Some officers of HMS King George V suffered a similar rude shock in an officers' mess at Piddlehinton. They had been invited to a regimental guest night of the Parachute Regiment stationed there shortly after the war. After dinner and a considerable amount of alcohol the hosts suggested as a party piece a 'quick stick' out of the window to the grass below. With immaculate efficiency the home team solemnly fell in behind the window and by use of a chair climbed on to the window ledge and on the command 'Green light on – Go' duly plunged out into the darkness at record speed and equally rapidly returned to the mess for another round of drinks amid great applause by the guests.

The Royal Navy, not to be outdone, prepared for their first para simulated descent, but in the meantime the mess staff had not been idle having moved a large water tank and placed it under the window. The naval squad returned wet through and well frozen as it was winter. Fine hospitality indeed, comments our contributor. Fine indeed! It wouldn't have done in the old days!

The same Tank Corps mess at Catterick were guilty one guest night of carrying out an unauthorised modification to a WD building. This particular mess, reportedly built by German POW in the first world war, was a pretty shoddy one-brick-thick affair. A corridor ran the length of the ante-room leading to the billiards room and other quarters beyond, and the dividing wall was not of the strongest. The mess had some unusually large and stalwart officers, who were pretty well matched by the guests, and during one concerted rush in a game of mess rugger or something similar the whole party crashed into the wall. Party and wall subsided in disorder. The DCRE, an understanding officer, decided to let matters rest. The debris was removed, all 'made good' and the ante-room vastly improved, but it was made clear to other regiments that no precedent had been established and that any proposed alterations to a building would have to be the subject of the customary Board of Officers.

The Essex Regiment claim a unique after-dinner diversion. The guest of honour was liable to find himself rolled up in the carpet and put aside to digest and ponder on the vagaries of life. This strange behaviour would also assist in setting the stage for rougher and stranger activities such as 'Highcockalorum and Jig'. In this very athletic pastime, two sides of ten are picked, mess jackets and spurs removed and, if wise, collars are undone. The side losing the toss (jig) form up one behind the other, with the front chap against the wall in position as in a rugger scrum (sometimes he stands facing into the room so that he can warn his team of impending events and help them repel boarders) and with the remaining nine lined up behind, each with his head and shoulders against the backside or between the legs of the man in front. Thus a firm bridge or jig is formed.

The players of the other side now in turn run as fast and far as the length of the room will permit and vault on to the jig, with the object of getting all their team on to the jig and breaking it down. If they do and the jig collapses they are the winners and are rewarded by the jug with their chosen alcoholic refreshment. If, owing to obstruction by the jig (and here the point of No.1 facing inwards can be seen) or due to faulty jumping, they fail to get all ten on, the jig win and are suitably rewarded. If all ten do get on but the jig does not collapse, it is a draw and everyone buys everyone else a drink – probably by the jig. A good tactic was to have a large officer built on Goering lines in your team and put him in at about fifth when a nice build up had been created and the jig might be starting to sag. His added weight should break the jig's back. But if he failed it would be very hard to get the remainder of the team on. So it was a nice decision to make, when to put him in. The success of this tactic also depends on the length of the available runway and speed of your jumbo jet or he may never become airborne!

A mess hardy perennial was cock fighting. In this the opponents lie down on their backs, shoulder to shoulder, legs stretched out in opposite directions and their right arms interlocked at the elbow. On the command to start each raises the right leg and after a bit of tentative sparring one will try to make a quick hook round the other's leg and with a downward twist turn him over. It will be seen at once that spurs cannot be worn and it was advisable to unstrap overalls and loosen braces to give more freedom to the leg and avoid splitting one's overalls.

At a regimental guest night in India Littlewart was matched against the Maharajah of Jaipur.

The latter was resplendent in the uniform of his state cavalry, topped – though that is scarcely the appropriate word – with gold spurs, which he removed and handed to his ADC who was little less resplendent. The first leg went to Littlewart easily and he had visions of receiving, even claiming those spurs as a souvenir of victory. The Maharajah had no such intentions. He proved to be as good at cock fighting as he was at polo, quickly made it leg all and after a prolonged bout of sparring won the final leg. Clanger was taken aside by his CO and, for the first and only time in his career, complimented on his tact; a gesture that he did not deserve as he had tried his hardest to win his spurs!

Another very popular diversion was 'Are you there Moriarty?'. Again the contestants lie on the floor, this time blindfolded and on their stomachs with left hands clasped at full arm's length. In their right hands they hold their weapons – a rolled up copy of the *Times* or perhaps its fellow traveller the *Pink 'Un.* The first to strike challenges 'Are you there Moriarty' to which without undue delay Moriarty has to reply, trying to throw his voice. The questioner then lets fly at where he imagines Moriarty's head might be. It is then Moriarty's turn. Basically very simple. A lot of the entertainment is caused by Moriarty's contortions to avoid being hit. He may twist and manoeuvre as he likes as long as his left elbow remains stationary in position. Some even turn over on their backs and risk a 'smack on the kisser'; others pull themselves right up over the left elbow and hope the blow may fall near the waist line but if an ambitious messing officer, recently returned from a holiday abroad, has put on the dish he liked so much at that little fishing village on the Costa Astronomica, the main component of which seemed to be garlic, this manoeuvre would be at once detected and a direct hit probably follow. The winner is the first to make three good hits or whatever is decided.

Of course, this game should only be played in a spirit of fun. When blood lust enters into it all entertainment value vanishes, except for the sadist. Clanger was once a victim when his opponent, one of his hosts, removed his blindfold exchanged his *Pink 'Un* for the *Illustrated London News* and belaboured him unmercifully until he rumbled he'd been taken for a ride. He never visited that mess again.

There is one game that some of the sadists enjoyed, which might be called Heidelberggefinger-bracken since it has an affinity with the duelling beloved by Heidelberg students. The two duellists stand at opposite ends of the billiards table; one hooks a finger over the cushion and the other tries to hit it with a billiard ball. To maintain direction, of course, the ball has to be propelled with considerable force and a little extra is added to increase the pleasure for the 'bowler'. Too much extra and direction is lost so there is a little skill in assessing the strength to give the maximum accuracy with the maximum pain on impact! Yet screeching with pain they seem to enjoy it! Unlike duelling, they have no 'honourable' scars to dazzle impressionable frauleins.

Having reached the billiards room let us consider some of the quieter but very entertaining games connected with it. In the old days the billiards table provided a great source of entertainment – as long as there were not too many aficionado in the mess and the table became as sacred as the table at Lords. Unfortunately there are nowadays not so many billiards rooms. They have been converted into ladies rooms, telly rooms or air-conditioned bars, but where they remain they can be the venue for some of these harmless games.

Billiards cricket is an excellent game calling for quite a lot of skill and control; accuracy of eye and hand are essential. The wicket is set up at the bottom spot where the two whites are placed a ball's width apart. The batsman wields his cue between these two balls; if he hits either he is out and if he raises the cue above the top of the balls he is also liable to be given out though he may receive a caution first. He is naturally out if the bowler hits either of the white balls or if he is caught in the field. The field is placed by the bowler just as for real cricket except that there is obviously no wicket keeper. If the batsman hits the ball on to one of the fielders he is out – caught – and the fielder is replaced but if the bowler hits a fielder that fielder is removed for the rest of the innings. The bowler bowls by rolling the ball out of the D at a reasonable speed; he will be no-balled for excessive speed and each member of the bowling side must bowl an over.

The batsman scores when he hits the ball to the top cushion (4), side cushions above the middle (3), below the middle (2) and the back cushion (1). Into a pocket can be the subject of local rules – either 6 or out. There are the usual two umpires, one at the bowlers' end to control the speed of the delivery and one at square leg to watch the height of the cue above the wicket. It is surprising what good scores can be made and how quickly a bad bowler can wreck the field.

There is a variation whereby the bowler bowls by rolling the ball down two cues held together with their points in the D. This can provide great accuracy or, in the hands of a novice, scatter the field even more quickly than by hand. But it savours rather of the Military College of Science and is not to be recommended!

Bowls is another game that has two versions, the more entertaining of which is the one which most nearly resembles the real game. The jack is a white ball and there are two sides of three, colours v reds. The first man to bowl sets up the jack by rolling it out of the D in any direction, but it must hit two cushions before coming to rest. The bowlers follow in alternate turn each wood having to hit two cushions or be removed. An opponent's wood or the jack can be struck but only if two cushions have first been hit; failure entails loss of the wood and the replacement of those that have been disturbed. The aim is to get as many woods as possible lying closer than the opponents'. If in an unsuccessful attempt to disrupt the field the wood comes back over the baulk line it is removed.

In the other form of bowls three quadrants are drawn in chalk round a bottom pocket and given scores increasing in value as they get nearer the pocket. The aim is to get the highest score. A ball going into the pocket is lost and there is therefore not so much scope for bumping, so the game tends to be more boring.

There is no need to describe billiards or snooker – a game incidentally invented by an army officer in India – but the popular game of 'Slosh' if not already familiar from Sandhurst days might be described. It is, I believe, a variation of Russian Pool probably invented at the RMC.

The colours are set up as for snooker with a red ball added hard against the back cushion immediately behind the black. The colours carry snooker values and the red ball twenty, but to earn a score the player must go in-off or pot in the correct pocket. These are, yellow in the top right, green in the top left, brown in either top pocket; blue in the middle pockets only, black in the bottom pockets only, pink and red anywhere. The player may make a cannon against any two balls or score by potting or going in-off as noted but if in the course of a break he goes in-off, or if for any reason a colour falls in the wrong pocket, the break is forfeited. If the red falls and the player has not scored, twenty is deducted from his score, but if the red or a colour falls into an appropriate pocket the value is added to the break. Colossal scores can be made and as easily forfeited. It is a consoling game for the player who finds straightforward billiards or snooker too much. It also provides plenty of scope for local rules and variations.

It has probably been quite unnecessary to describe these games for most readers, but the accounts may prove of value to a regiment who, after years of sojourn in the wilderness, finds itself the lucky owner of a billiards table and does not know what to do with it.

The 12th Royal Lancers found an unusual use for a table when stationed in Malaya. Bull frogs abound in that country and were often rounded up for race meetings on the billiards table.

Fred Burnaby, who has already received a mention, could vault across a billiards table. He was immensely strong and a favourite trick of his was to hold a cue at arm's length, butt in the air, and the tip between the first two fingers of his hand. On one occasion he even twisted a poker round the neck of the Prince of Wales (later King Edward VII). But these diversions are only for the exceptionally talented.

If there was a piano in the mess and a pianist prepared to play all night – at pistol point if need be – there would probably be a bit of singing. The programme would include such songs as *She was poor but she was honest, If I were a marrying girl,* and with words and actions or actions only *The king was in his counting house, Oh, Sir Jasper* and many more. If the guest night was in a RA mess there would be the Gunner hymn, Kipling's *The Screw Guns* sung to the tune of the Eton boating song. In a Sapper mess, a crocodile round the building singing *The CRE* would probably bring the evening to a close.

xi
Nights of gladness

There are many nights which for no particular reason 'strike fire' and next day the wine sheets[1] come as rather a shock. These are nights of light-hearted gladness but heavy-headed sadness in the morning which is only dispelled by healing draughts of brandy and ginger ale, or black velvet. Nothing like black velvet to lighten the gloom! Such a morning after must have followed the first Candelabrum night in the Royal Artillery mess at Woolwich. The King, William IV, had presented a handsome candelabrum and was entertained to dinner so that the officers might express their appreciation. It must have been quite a night; each officer was charged three days pay, which they may not have appreciated in the cold, sober light of day! However, not every regiment has the opportunity to entertain Royalty in the mess and it must have been worth it.[2] Nowadays, the big nights are usually connected with celebration of the regimental day when a famous battle honour is recalled. They can be forecast months ahead and the wise officer may build up a sinking fund towards the night. Such a night might be on Sobraon Day when the Lincolnshire Regiment celebrate their most famous battle honour won on 10th February 1846. At dinner on the anniversary of that day the brigade commander's letter is, in accordance with his directions, placed on the mess table in a place of honour. The letter, addressed to Lt-Colonel Franks CB, who was CO at the time reads; 'I write to say that the gallantry of HM's 10th, and you, their gallant leader will never be effaced from my memory. The glorious conduct of the 10th at Sobraon is beyond any praise that I could give. They were the corner-stone of the victory. I have seen much service, but I never saw anything to equal their cool and resolute courage on that day. You will do me the favour to offer my explanations, how I left without calling, and allow this, my testimony of the steadiness and gallantry of the corps to be placed on the mess table'.

Minden day is celebrated by six regiments, the occasion when Thomas Carlyle described 'these unsurpassable Six, in industrious valour, unsurpassable'. On the anniversary they wear roses in their headdresses, but the Lancashire Fusiliers (20th Foot) go further. After dinner in the officers' mess every officer who has not previously done so stands up and eats a rose, which is handed to him in a silver finger bowl containing champagne. Until he has done so he is not considered to be a full-blooded Lancashire Fusilier or to have in him the spirit of Kingsley's men. (Kingsley was CO at the time of the battle.) A special toast is also drunk – the Minden Toast. The CO proposes the toast in these words – 'We will now rise and drink in solemn silence to those who fought and fell at Minden'.

1. In those days the chit system was not in force. Officers' wine sheets (i.e. their daily accounts) were made available daily for inspection, placed on a convenient table in the mess. If an officer thought he had been incorrectly charged for an item he had to bring it to notice within 48 hours. If he did not it was assumed that the entries for that day were correct and no alteration would be made. It was no use four days after a guest night claiming 'Oh but I can't have drunk that much'. You must have; your wine sheet said so and that was that!

2. Worthy of note might be the following record of consumption of wines in the RA mess Woolwich in 1827 on an occasion similar to a candelabrum night. Some 157 sat down to dine at a cost of 14s.0d. a head and despatched the following wines:

Claret	bottles	130	Sherry	bottles	59
Sauternes	"	51	Port	"	37
Marsala	"	42	Madeira	"	42

Heaven knows what the cost would be today – several hundred pounds probably.

At the end of dinner the Band march round the table playing the Minden March. As they leave the room the door is closed on the bass drummer so that he cannot 'escape' and he then makes a short speech, usually of a humorous nature (I dare say ribald!), drinks the health of the officers and departs. After such a start the evening must have developed on rather more than traditional lines.

* * * * * * * * *

A close relationship has existed between the Queen's and the Royal Navy since the Glorious 1st of June 1794 and every year joint celebrations take place in the Royal Naval Gunnery School at Whale Island and the barracks of the regiment, wherever they may be stationed, in alternate years. When the event takes place at Whale Island the highlight of the evening is the Whale Island gun drill competition, based on the interport gun drill competition which is such a popular feature of the Royal Tournament.

Teams of six are selected from the Queen's regiment and the naval officers and each team is issued with one bed GS, all ranks for the discomfort of, instead of a gun. This bed is the old fashioned type with two cast iron end pieces into the slots in which fits the wire spring frame for the mattress or army 'biscuits'. At the start of the competition the bed is in pieces and at the drop of a flag each team has to assemble its bed and place on board one officer. Then with that unfortunate clinging to the springs the teams run the length of the ward room. The winner is the team reaching the end of the room first. This sounds plain sailing, but there is the small problem of navigating an archway on the course. This entails tilting the bed at 45 degrees or the bed won't go through. If the passenger is clinging to the sides of the bed he will undoubtedly lose the skin off his knuckles and possibly his hold and fall in the drink; and that means back to the start and begin again! On the other hand it requires considerable strength to hook on to the springs. The passenger's lot is not a happy one.

The losing team buy beer for the winners and vice-versa ad infinitum or dawn – it is very thirsty work! It would make a nice change to put this contest on as a turn at Earl's Court.

A regimental guest night which developed rather unusually is recounted by an officer in the Devonshire Regiment.[3]

'In Calcutta the 1st Bn was stationed in Fort William, Kitchener's abode at one time, I believe. The bishop of Calcutta, a tall ascetic-looking bachelor with a twinkle in his eye, and a prelate of high renown, was a not infrequent visitor to the mess of which he was an honorary member. His chaplain was Tommy Tucker, who had been chaplain to the battalion in Quetta.

After my term as adjutant, having taken my young family to the UK, I lived in the mess but had no particular duties with the battalion except that I was wines' officer on the mess committee.

A regimental guest night was arranged during the hot weather. The bishop and his chaplain were to be present among the fifty or so dining. I had noticed on previous occasions that the bishop had a sherry before dinner, one or two glasses of wine or champagne, and that he evidently enjoyed his two glasses of the Boys' Port, so called. Here I must digress to explain that on a guest night it was customary to circulate this in the delightful ornate silver limbered wagon coasters, No 1 Port and the Boys' Port (cheaper) with a brown sherry in a lone coaster. As I have already said, it was extremely warm and a day or so before the due date, having consulted various wine connoisseurs as to what wines should be served, the mess sergeant, Tucker and I with the dear old Abdar hovering in the background, came upon the cases of port in the wine store. The bottles were distinctly warm to touch. I thought to myself, being partial to several glasses of port after a good dinner – and after discussion with the mess sergeant – that it would be an error, nay a sin to circulate port at a temperature higher than 60° F. So we decided to put a couple of cases in the cold room. This was a rather dark, dungeon-like pantry, about 14ft by 10ft with a heavily barred glassless aperture and fitted with a slate shelf running the entire length of the dungeon. The temperature was just over 70° F, but by dint of regular application of cold cloths we got the temperature of the bottles gradually down to about 60° F.

3. Major W.P. Symonds.

Came the dinner. Dress was blue overalls, white jackets, stiff collars and scarlet cummerbund. On guest nights I sat at 'cad's corner'[4] on Mr.Vice's right so that I could indicate to him any hiatus in service or instruction necessary to the mess staff. Apart from the adjutant I was the senior subaltern. I could observe the colonel and the bishop from my seat and also have a 'backhander' or two from Mr.Vice!

The bishop was enjoying himself: he had two refills to his champagne, his glass of port for the loyal toast, and, it being a very good dinner, good company at table and cheerful conversation, he showed no signs of wanting to leave and, moreover, had two more glasses of port.

In due course we rose from the table and congregated on the verandah – a delightful place with room for many people to assemble and with the large bladed fans manfully circulating slightly cooler air. Round the piano, at which Tommy Tucker was playing soft music – he was a gifted FRCO – were seated the CO, the bishop and some seven or eight senior officers and guests enjoying their coffee liqueurs and cigars. As I passed by on my way to join a less exalted group I heard the bishop call me. "Simpkins" (my nickname for many years), he said, "What was the excellent port I was enjoying so much tonight?". Smiling to myself with all eyes on me I replied, "The port you usually enjoy sir". "Well," he retorted, "I must confess that I came to the conclusion that it was something special so I had rather more than I usually do." Or words to that effect. I forget who chipped in with "Are you sure Simpkins?". "Yes", I replied, "as a matter of fact the weather is so hot and the bottles felt so warm to the touch a day or so ago that the mess sergeant and I decided to put a couple of dozen bottles or port in the 'cold room'."

Dead silence, then a murmur or two evidently of disbelief. As I was about to move on the bishop spoke again.

"Colonel", he said, "What is the penalty to be imposed on an officer who ices port wine?" (I remember well that he added the word 'wine'). More murmurs and chuckles. "I can't think of any punishment severe enough" says the colonel, "but as you were the one victimised and as honoured guest bishop, you must pass sentence." Taking the cigar from his mouth and looking most distinguished in his thin purple silk cassock, silver chain and cross, the bishop thought for a moment, then with a twinkle in his eye said, "I think a measure of liqueur brandy all round would meet the case and we might include Simpkins to console him".

Again murmurs, this time of amused approval and anticipation. The old Abdar stepped forward and signalled a hovering Khidmutgar to whom the order for nine (I think it was) brandies was about to be given. Deciding to get a little of my own back and not being partial to brandy, I put on what I hoped was suitably ashamed but at the same time defiant look and exclaimed, "Being only Clifton and Sandhurst, where I may say with due humility I was an under officer, I paid more attention to my studies than to the cultivation of my palate with expensive tastes, so I will settle for a large Benedictine laced with my brandy".

Smiles, snide remarks and laughter and the order was given. Next morning I checked my wine-sheet. My Lord bishop, bless his heart, had had these drinks put down on his sheet.'

Not always does a summary award in the mess have such a happy ending Simpkins was lucky to get away with it. The evening which he has described had no particular excuse but was one which clearly has been impressed indelibly on his memory and it does give a picture of the very happy nights in mess and in this case a very happy morning too.

* * * * * * * * *

Another night which must have lived in the memory of all who were present would have been the combined dinner of both battalions of the Seaforth Highlanders in Palestine in 1934 attended

4. Also known as 'starvation corner' as the unlucky occupant of that place was last to be served.

by wives. Not only was it the first time that ladies had graced the board but for both battalions to be in the same 'station' and abroad at that would have been almost certainly unique. The invariable practice in those days was for one battalion to be at home and the other on foreign service. The remarkable coincidence enjoyed by the Seaforths was brought about by the Arab 'troubles' in Palestine at that time. Nowadays such a situation would no longer be possible – with infantry regiments reduced to one battalion and that one an amalgamation. Although the 1st and 2nd Bns Royal Anglian regiment might conceivably find themselves lying alongside one another it would in no way be comparable with the unusual and happy situation the Seaforths found themselves in.

* * * * * * * * *

National Saints' days are 'naturals' for nights of gladness and probably pride of place must be taken by St.Andrew's day. The English do little to celebrate St.George's day – the Irish often spend a week – three days before and three days after to make sure of the date – to celebrate St.Patrick's, but the piping in of the haggis has a place all of its own with St.David only as a close runner up, with the eating of the leek, look you. But then, whilst all Scottish regiments will pipe in and hog the haggis not all Welsh regiments will observe a leek-eating ceremony. So let the Scots take first knock and describe the procedure for St.Andrew's night in the mess.

'In accordance with the old Scots Fusilier custom, the procedure for ordinary guest night holds good except that the haggis is served as an entree after the fish course. The haggis is piped in by the Pipe-Major and whisky is served with it. The Pipe-Major will lead followed by the cook carrying the haggis above his head, with the mess sergeant bringing up the rear with the commanding officer's dirk on a salver. The party will move in a clockwise direction and complete one full circuit of the table from the commanding officer's seat. After completing the circuit the party will halt and turn inwards with the pipe-major immediately behind the commanding officer. As the piping ends the mess sergeant will step forward and offer his dirk to the commanding officer. When the commanding officer has taken it the cook will step forward and the commanding officer will pierce the haggis. The cook will then step back, the mess sergeant will take the dirk from the commanding officer and in turn step back into line. This will be the signal for the party to turn to the left and step off. The procession will leave the mess again completing one full circuit from the commanding officer's chair before leaving the dining room.

The piper will play the haggis in and out to *Brose and Brochan.*

While the haggis is being eaten the quaichs are handed round. Two quaichs will be carried on each salver, one for whisky and one for athole brose. One quaich is handed to the padre (if present) and the other to the officer diagonally opposite to him across the table. In the absence of the padre the quaichs start with the president and the vice-president. They are filled with whisky and athole brose. Both officers rise and facing each other propose 'The Pious Memory'. They drink the contents of the quaich, turn it in one continuous movement, and kiss the bottom. The quaichs are passed round the table each person repeating the ritual. Athole brose is passed round with the port, and the mull with snuff handed round with the cigarettes.

At the beginning of the pipe programme, the tune *St.Andrew's Cross* will be played as a separate piece, the pipers leaving the dining room after playing it.'

That is the procedure as laid down in the mess rules of the Royal Highland Fusiliers. What follows, of course, is governed by no rules.

Although eating haggis may daunt many a Sassenach it is part of the meal. The thought of having to eat a raw leek *after* having enjoyed a good meal washed down with wines to match is positively forbidding, moreover it is a prospect facing not only the newly-joined officer of the Royal Welch Fusiliers, but also the guests at a dinner when the ceremony is to be carried out.

The Regimental ceremonial of eating the leek has remained almost unchanged over the years. It was described in detail in the records as having taken place at Boston, Virginia, in 1775 and referred to then as an ancient custom. The goat is led round the table followed by a drummer, fifers, the drum major carrying a silver salver on which are the leeks, and the mess sergeant with a loving cup. They halt by the latest joined subaltern, who stands on his chair and with his right foot placed on the table he is offered a leek. While he eats the leek the drummer plays a continuous roll until it is completely consumed. The mess sergeant then hands him the loving cup for a toast to St.David. But whilst this is taking place let us make the closer acquaintance of the goat – never to be referred to as a mascot, please!

Though no record exists of the origin of the custom that the Royal Welch Fusiliers march with a goat with gilded horns at their head, it was apparently a custom of some long standing in 1777. Furthermore the goat is known to have been in action at the Battle of Bunker's Hill in 1775. The first Royal goat was presented by Queen Victoria in 1884 and since then the regular battalions and most of the territorial battalions have had a goat presented by the sovereign. The goats are selected from the royal herd which was started from a pair given to Queen Victoria by the Shah of Persia. In recent times when no goat was available from the royal herd, Her Majesty the Queen has been graciously pleased to present a wild goat from the mountains of North Wales where several herds exist.

The goat is an essential part of the regiment and wherever the regiment goes there also goes the goat. Many a tale can be told of the efforts to find replacements when the goat has been a casualty whilst the regiment was overseas. In the Crimea the goat died and great was everyone's surprise and delight when one appeared as usual on St.David's Day in the officers' mess. With his horns gilded and wearing his silver head plate he was led around the table but on reaching the colonel's side he stood on his hind legs and drank a toast to St.David. This loyal goat was none other than Private Styles, the colonel's groom, whose long Service Medal can be seen in the Crimea Case in the regimental museum.

Each regimental goat has his own silver head plate, and except for the silver breast plate worn by the goat of the 1st Battalion, wears no other decoration. He is always in the care of a drummer who holds the honorary title of 'Goat-Major'.

By now 2/Lieutenant Shenkin should have eaten his leek so let us return to the dining room. When the newest-joined subaltern has eaten his leek the procession goes to each officer of the regiment who has never previously eaten his leek with the regiment and the ceremony is repeated. It is then the turn of the guests who are invited to partake likewise; when all have eaten their leeks the band plays *Men of Harlech* and the procession marches round the dining room and out, and the ceremony is over.

As already noted, the Royal Welch Fusiliers only drink the loyal toast on St.David's Day and included on that day is the toast to 'Toby Purcell, His Spurs and St.David'. Major Tobias Purcell was the first second-in-command of the regiment. He distinguished himself at the Battle of the Boyne and on the death of Colonel Herbert succeeded to the command. The spurs which he wore were preserved in the regiment and kept in the custody of the second-in-command. To prove that they were safe they were produced each St.David's Day and the toast drunk to the music of *Jennie Jones.* Unfortunately the house of Major Holmes, then 2IC, was burnt down in Montreal in 1846 and the spurs destroyed. But the toast, proposed by the second-in-command of the time, is still drunk.

There is no comparable ceremony on St.Patrick's Day in the army. The 'Micks' do not eat a raw potato washed down with Irish Coffee, in any event a potato would probably be a too unhappy reminder of days 'jankers' spud peeling, but there is the ceremony both practised and known all over the world of the distribution of the shamrock to the Irish Guards. This takes place in the morning and the rest of the day is observed as a holiday in such ways as appeal to all 'Micks'. Whilst St.Patrick is reputed to have expelled all snakes from Ireland, on his day some of them make an unwelcome return to those 'with the vision'.

* * * * * * * * *

The RA mess at Woolwich has built up since its earliest days a reputation for grand and glorious entertainment and it is, therefore, fitting that this chapter should end with a description of a big gunner night.

The Spring Dinner is the most important of the regimental guest nights in the Royal Artillery mess, Woolwich. The Master Gunner presides and about ten regimental guests are invited, drawn from the upper echelons of politics and public life, the services, judiciary and commerce. Officers of the regiment and their private guests make up the remainder of those attending up to the capacity of the mess, namely – 134.

Officers' Mess, Cavalry Depot at Canterbury. 1890's.

Tables set for The Spring Dinner - R.A. Mess, Woolwich.

When a Mess Ball was held, and ladies invited, 'Kala Juggahs' or Dark Places were carefully arranged by the Mess servants on the verandahs for (as 'Hobson-Jobson' stated in 1886) 'the purpose of flirtation'.

See page 119.

'Number One' brings in the 'Shroff's' bills on a Guest Night.

See page 119.

At about 7.30 pm. all assemble in the ante-room the officers wearing mess dress and retired officers and civilians in tailcoats. The regimental guests are escorted in to meet the Master Gunner, the Director of Artillery, the colonel's commandant and other senior officers who will be their hosts at dinner, and sherry is served. A relaxed half-hour follows with officers and their guests moving round the room to greet and chat with old friends. At 8 pm. the mess steward appears in the door and announces 'Gentlemen, dinner is served'. Thereupon four trumpeters of the King's Troop RHA, wearing full dress and stationed in the hall, sound the mess call.

After a suitable interval the Master Gunner and the guest of honour lead the procession into the mess room. This magnificent room never fails to impress those who see it for the first or indeed the hundredth time. Very large – ninety feet by forty five – but beautifully proportioned, it is maintained very much as it was in Georgian days. The six great chandeliers, the royal portraits on the walls and the graceful Georgian mirrors form a perfect setting for the four great mahogany dining tables, now laid out in all their guest night finery. Three of these tables run the length of the room while the fourth, the top table, is placed across the room at one end. All are decorated with the famous Woolwich Silver collection; the whole illuminated by the soft light of about a hundred candles in their silver candelabra.

The Master Gunner takes his place in the middle of the top table and the officers and guests file in to their appointed seats, while the string orchestra of the Royal Artillery band plays *The Roast Beef of Old England* from the far end of the room. When all are standing behind their chairs the Master Gunner asks the padre to say Grace, or says it himself if no padre is present. The regimental Grace, formerly used in the old Royal Military Academy at Woolwich, is appropriately short and sharp – 'For what we are about to receive, Thank God'.

All now take their seats in the large and comfortable hundred-year-old dining chairs and dinner gets under way, served by the entire mess staff, an expert and well drilled body of men and women. Some of the waiters wear the old Georgian livery of the mess – long cream coat, knee breeches and powdered wig. The WRAC waitresses wear their special dinner uniform – a blue jacket with scarlet lapels to match the Gunner mess kit, and blue skirt. Appropriate wines are served with the various courses – sherry with the soup, hock with the fish, the mellow mess burgundy or claret with the main course and champagne with the sweet. The hubbub of conversation becomes more animated as the leisurely meal proceeds, almost drowning the light music played by the orchestra.

After the last course the tables are quickly cleared of china, cutlery, glasses and linen. The expertly executed manoeuvre for the long runners always gets a round of applause. A waiter and a waitress take post at the ends of each thirty-six foot runner and twist it into a coiled spring. Then, at a signal from the mess steward each waiter gives a deft heave and the long runners slide down the tables and over the end in one swift and graceful movement.

The port glasses are now laid out and the decanters of port and madeira make their way up and down the tables. When all glasses are charged the mess steward reports 'Wine passed sir' to the Master Gunner who raps sharply with his gavel. Conversation is immediately hushed and all rise for the loyal toast, the Gunner officers holding their glasses stiffly in front of them according to regimental custom.[5] The Master Gunner proposes the toast in the words 'Gentlemen the Queen – our Captain-General' and the orchestra breaks into the National Anthem. At the same time the hundreds of lamps in the chandeliers are switched on flooding the room with a bright light, strangely harsh after the soft glow of the candles. The National Anthem ended, all say 'The Queen' and drink the toast. After a small silent moment the chandeliers are switched off and all resume their seats and their conversation. At Gunner regimental dinners there are no further toasts and no speeches, to the great relief of hosts and guests alike. The port circulates again, cigars are served and lit and the snuff boxes are passed round with ritual tapping. Coffee follows and balloon glasses of brandy. The band now presents its famous rendering of the *Post Horn Gallop.* Two expert musicians (never bandsmen in the Royal Artillery) take post in diagonally opposite corners of the great room and perform first on the long coach horn and then on the shorter and livelier post horn, accompanied by the orchestra with a stirring arrangement of *John Peel.* Thunderous applause greets this turn and an encore is given. The Director of Music meanwhile enjoys a well deserved glass of port with the Master Gunner.

5. As described in Chapter V.

And now the dinner is drawing to its close and the band starts to play the marches – first the march-past of each of the regimental guests who has served in the armed services and then the four regimental marches of the Royal Artillery – *The British Grenadiers, The Keel Row, Bonnie Dundee* – and the tune loved by all Gunners *The Royal Artillery Slow March.* After the marches the garish chandeliers are again switched on and the Master Gunner leads the guests out of the mess room. The company now starts to break up, though many linger over their glasses of brandy and walk around admiring the mess silver – the great centre-piece, each with a facet of regimental history attached, the grand memorial to a Field-Marshal and the poignant memorial to a subaltern; the gold cups given by sovereigns and princes and the many exquisite pieces presented by other regiments.

Now all have gravitated to the ante-room and the bar, where brandies and sodas are consumed – loud but decorous conversation prevails. At length, around midnight, the coach leaves with its passengers for London and those driving themselves home call for prairie oysters before cautiously departing. The mess is now deserted except for the wise individuals who have booked rooms for the night.

The Spring Dinner is over.

xii

Major-General Littlewart is dined out

Major-General C.O. Littlewart CB, CBE, DSO (even his rivals admitted that the honours had been well earned) sighed as he pulled on his overalls. It might be the last time he would wear mess kit as a serving officer. Tonight he was dining with the officers of the Airborne Forces depot in Aldershot. In recognition of the time he had spent airborne some years ago before becoming, as he put it, chairborne they were dining him out in their mess. He had, of course, been dined out by the Ramrods when he finished his time as CO for most officers being dined out of their regiment meant out of the army as well – not many were lucky enough to go further. He had been a popular and, let it be said, far from inefficient CO. Rather the reverse although, possibly on the slow side and better at Q than G.[1]

His dining out had been hilarious. He had defeated all efforts to 'nobble' him; had romped through Cardinal Puff faultlessly; had cock-fought and defeated the newest joined 2/Lt.from the RMAS, a hefty rugger blue who Clanger thought was a cocky young bastard and required to be taken down a peg. He was. He had conducted the band[2] and had even attempted to play the cornet – he had been a bugler in his OTC band – much to the delight of the bandsmen but to the apprehension of an over-anxious bandmaster; in fact the night had been a triumph.

It was a custom in the Ramrods that the officer being dined out should be conveyed to his quarter – if the distance was not too excessive – seated on a pot, chamber, china, crested (for senior officers only) placed on a six foot table which was carried by as many officers as thought fit and were fit – or thought they were. Rather like being chaired at Bisley but considerably more hazardous.

It was a form of poetic justice that Clanger, who had dropped so many in his time, should on this occasion be himself dropped, bang in the middle of the square on the top of his head. He promptly lapsed into the state of unconsciousness that all the ministrations of his officers and all the potations he had drunk had failed to achieve. A stretcher was hastily summoned from the Guard Room and on this he was reverently carried to his quarter and laid in state on a settee in his sitting room. There a little while later the Medical Officer pronounced him to be in no danger and there was no need for arms reversed. This did not surprise some officers, those present at the scene and those absent. They unkindly considered that he had been wounded in his least vulnerable spot. His detractors insisted that the fall had knocked the last vestige of sense out of his head and had so guaranteed his further advancement. On the other hand his admirers asserted that it only gave further proof of how tough he was. If there was anyone, they said, who would make a parachute jump without a parachute it was old Clanger. A statement that was open to more interpretations than one – not all flattering!

1. At the Staff College, Littlewart had learned the DS (Directing Staff) solution of the difference between G and Q work. G it was said was like playing golf without the ball – easy; but Q was playing golf with the ball which was a damned sight harder. Nevertheless he liked Q and found he had a distinct bent that way, and this largely contributed towards his advancement. Officers who *like* Q are not very numerous.

2. In the Border Regiment it was a custom that a CO on being dined out was invited to conduct the band as it played the regimental march.

Nostalgia

Maj-Gen. Littlewart on washing up fatigue in clouds of steam and pretty bubbles from 'Fairy Liquid' wistfully dreams of the good old days in the Mess. Then there were others to carry out the plates and wash up and they did not have to worry about grandchildren's 'potties'.

When approached about it himself he said only 'Fall? what fall? Can't remember a thing — damned fine night. Bit of a head in the morning of course but that's nothing unusual and a hair of the dog[3] soon cured that'.

Now as he carefully brushed his hair over the old scar he wondered what might be in store for him tonight. He should have known; or guessed. He should have remembered that history has a habit of repeating itself. It did. It was a glorious night. He more than held his own in the athletic pursuits; refused to be caught perjured at liar dice; won a lot at roulette and all in all showed the younger generation that there was still plenty of life left in the old dog. The younger generation began to think that there was a damned sight too much, in fact in the words of the song they did not want to lose him but thought he ought to go. Not that he had over-stayed his welcome but he was getting too expensive, the old chap never lost! When the time came for him to make his farewell it was unanimously decided that he must leave in a 'stick' and one was lined up. The General should have seen the writing on the wall or recalled the crest on the chamber pot but of course he did not and lined up with the rest with the greatest alacrity. Last to leave the aircraft he plunged to earth. He landed awkwardly, there was a loud crack and in a very short while he was on his way to the Cambridge Military Hospital where for the first and last time in his life he had to be cut out of his wellingtons[4] and overalls and be put to bed — with a broken leg.

There in plaster we will leave him determined this time that he really will write his memoirs. I do hope he does,

..... BUT .....

3. In the last century Ensign Sir Morgan Odoherty of the 44th Foot recommended as a pick-me-up 'five drops of laudanum and a teaspoonful of vinegar in a tumbler of spring water'. Littlewart, like many of us, was a firm believer in the medicinal qualities of a brandy and ginger to dispel the weight and gloom of a hangover.

4. Littlewart heard many tales of officers who had to be cut out of their wellingtons because they were too port-logged to take them off but he had never encountered an actual case. Wellingtons were expensive; there would probably be little left of the night for sleep; for himself he would have waited for the arrival of his batman in the morning and let him tackle the job of extrication!

xiii
Amalgamations and Regimental nicknames

AMALGAMATIONS

The Army during its history has undergone many reforms and re-organisations which have usually resulted in disbandments, reductions or amalgamations. The amalgamations of cavalry regiments are mostly of more ancient origin than those affecting the infantry and are probably familiar to most people. In the infantry the changes over the years, which have culminated in the present day amalgamations, can be traced first to the reorganisation of 1881 when the numerical titles of regiments were dropped and county titles introduced and two regiments of Foot combined to form the 1st and 2nd Battalions of a county regiment.

The next step was taken after the last war when the county regiments were reduced to a single regular battalion and then many of these battalions were in turn amalgamated into big regiments of three or more battalions. Possibly by 1981 there will be further reductions and the few remaining infantry regiments may revert to numerical titles and a Corps of Infantry!

Below is set out as a simple 'form at a glance' the list of these amalgamations which may assist towards a better understanding of Chapter V and may be also of general interest.

CAVALRY AMALGAMATIONS

Original Title	Amalgamation	Present Title
1st Dragoon Guards 2nd Dragoon Guards	1st The Queen's Dragoon Guards	1st The Queen's Dragoon Guards
4th Dragoon Guards 7th Dragoon Guards	4th/7th Dragoon Guards	4th/7th Royal Dragoon Guards
5th Dragoon Guards 6th Dragoons	5th Dragoon Guards	5th Royal Inniskilling Dragoon Guards
13th Hussars 18th Hussars	13th/18th Hussars	13th/18th Royal Hussars (Queen Mary's Own)
14th Hussars 20th Hussars	14th/20th Hussars	14th/20th King's Hussars
15th Hussars 19th Hussars	15th/19th Hussars	15th/19th The King's Royal Hussars
5th Lancers 16th Lancers	16th/5th Lancers	16th/5th The Queen's Royal Lancers
17th Lancers 21st Lancers	17th/21st Lancers	17th/21st Lancers

These amalgamations took place in 1920-22.

Cavalry Amalgamations - continued.

3rd Hussars 7th Hussars	The Queen's Own Hussars	The Queen's Own Hussars
4th Hussars 8th Hussars	The Queen's Royal Irish Hussars	The Queen's Royal Irish Hussars
9th Lancers 12th Lancers	9th/12th Royal Lancers (Prince of Wales's Own)	9th/12th Royal Lancers (Prince of Wales's Own)

These amalgamations took place in 1958-60.

The Royal Horse Guards 1st Royal Dragoons	Amalgamated in 1960 to form The Blues and Royals.

INFANTRY AMALGAMATIONS

Pre-1881 Regt.of Foot	County title after 1881	Present amalgamated title
1st	The Royal Scots (The Royal Regiment)	—
2nd	The Queen's Royal Regiment (West Surrey)	The Queen's Regiment
3rd	The Buffs (Royal East Kent Regiment)	The Queen's Regiment
4th	The King's Own Royal Regiment (Lancaster)	The King's Own Royal Border Regiment
5th	The Royal Northumberland Fusiliers	The Royal Regiment of Fusiliers
6th	The Royal Warwickshire Regiment	The Royal Regiment of Fusiliers
7th	The Royal Fusiliers (City of London Regiment)	The Royal Regiment of Fusiliers
8th	The King's Regiment (Liverpool)	The King's Regiment
9th	The Norfolk Regiment	The Royal Anglian Regiment
10th	The Royal Lincolnshire Regiment	The Royal Anglian Regiment
11th	The Devonshire Regiment	The Devonshire and Dorset Regiment
12th	The Suffolk Regiment	The Royal Anglian Regiment
13th	The Somerset Light Infantry (Prince Albert's)	The Light Infantry
14th	The West Yorkshire Regiment (The Prince of Wales's Own)	The Prince of Wales's Own Regiment of Yorkshire
15th	The East Yorkshire Regiment (The Duke of York's Own)	The Prince of Wales's Own Regiment of Yorkshire
16th	The Bedfordshire and Hertfordshire Regiment	The Royal Anglian Regiment
17th	The Royal Leicestershire Regiment	The Royal Anglian Regiment (Independent 'Tiger' Company)
18th	The Royal Irish Regiment	Disbanded in 1922
19th	The Green Howards (Alexandra Princess of Wales's Own Yorkshire Regiment)	—
20th	The Lancashire Fusiliers	The Royal Regiment of Fusiliers
21st	The Royal Scots Fusiliers	The Royal Highland Fusiliers (Princess Margaret's Own Glasgow and Ayrshire Regiment)

22nd	The Cheshire Regiment	—
23rd	The Royal Welch Fusiliers	—
24th	The South Wales Borderers	The Royal Regiment of Wales
25th	The King's Own Scottish Borderers	—
26th 90th	The Cameronians (Scottish Rifles)	—
27th 108th	The Royal Inniskilling Fusiliers	The Royal Irish Rangers (27th Inniskilling 83rd and 87th)
28th 61st	The Gloucestershire Regiment	—
29th 36th	The Worcestershire Regiment	The Worcestershire and Sherwood Foresters Regiment
30th 59th	The East Lancashire Regiment	The Queen's Lancashire Regiment
31st 70th	The East Surrey Regiment	The Queen's Regiment
32nd 46th	The Duke of Cornwall's Light Infantry	The Light Infantry
33rd 76th	The Duke of Wellington's Regiment (West Riding)	—
34th 55th	The Border Regiment	The King's Own Royal Border Regiment
35th 107th	The Royal Sussex Regiment	The Queen's Regiment.
37th 67th	The Royal Hampshire Regiment	—
38th 80th	The South Staffordshire Regiment	The Staffordshire Regiment (The Prince of Wales's)
39th 54th	The Dorsetshire Regiment (Since 1951 The Dorset Regt.)	The Devonshire and Dorset Regiment
40th 82nd	The South Lancashire Regt. (The Prince of Wales's Vol.)	The Queen's Lancashire Regiment
41st 69th	The Welch Regiment	The Royal Regiment of Wales
42nd 73rd	The Black Watch (Royal Highland Regiment)	—
43rd 52nd	The Oxfordshire and Bucking- hamshire Light Infantry	The Royal Green Jackets
44th 56th	The Essex Regiment	The Royal Anglian Regiment
45th 95th	The Sherwood Foresters (Nottinghamshire and Derbyshire Regiment)	The Worcestershire and Sherwood Foresters Regiment
47th 81st	The Loyal Regiment (North Lancashire)	The Queen's Lancashire Regiment
48th 58th	The Northamptonshire Regiment	The Royal Anglian Regiment
49th 66th	The Royal Berkshire Regiment (Princess Charlotte of Wales's)	The Duke of Edinburgh's Royal Regiment (Berkshire and Wiltshire)
50th 99th	The Queen's Own Royal West Kent Regiment	The Queen's Regiment
51st 105th	The King's Own Yorkshire Light Infantry	The Light Infantry
53rd 85th	The King's Shropshire Light Infantry	The Light Infantry

Infantry Amalgamations - continued.

57th 77th	The Middlesex Regiment (Duke of Cambridge's Own)	The Queen's Regiment
60th	The King's Royal Rifle Corps	The Royal Green Jackets
62nd 99th	The Wiltshire Regiment (Duke of Edinburgh's)	The Duke of Edinburgh's Royal Regiment (Berkshire and Wiltshire)
63rd 96th	The Manchester Regiment	The King's Regiment
64th 98th	The North Staffordshire Regiment (The Prince of Wales's)	The Staffordshire Regiment (The Prince of Wales's)
65th 84th	The York and Lancaster Regiment	—
68th 106th	The Durham Light Infantry	The Light Infantry
71st 74th	The Highland Light Infantry (City of Glasgow Regiment)	The Royal Highland Fusiliers (Princess Margaret's Own Glasgow and Ayrshire Regiment)
72nd 78th	Seaforth Highlanders (Ross-shire Buffs, The Duke of Albany's)	Queen's Own Highlanders (Seaforth and Camerons)
79th	The Queens Own Cameron Highlanders	
75th 92nd	The Gordon Highlanders	—
83rd 86th	The Royal Ulster Rifles	The Royal Irish Rangers (27th Inniskilling 83rd and 87th)
87th 89th	The Royal Irish Fusiliers	The Royal Irish Rangers (27th Inniskilling 83rd and 87th)
88th 94th	The Connaught Rangers	Disbanded 1922
91st 93rd	The Argyll and Sutherland Highlanders (Princess Louise's)	—
	The Rifle Brigade. (Taken out of the line when the 95th Foot to form the Rifle Brigade in the early 19th century.)	The Royal Green Jackets

Regiments which have no amalgamated title shown against them escaped the massacre and retain their identity unchanged.

NICKNAMES

The British Army is rich in nicknames and many a regiment has become famous with the public by virtue of its sobriquet. The following list is based on one from 'Regimental Nicknames and Traditions of the British Army' — Gale & Polden, 5th Edition. Numbers of regiments and times of raising were complicated by the fact that they were numbered when they came on the English establishment, e.g. The Royal Scots Fusiliers raised in 1678 became the 21st Foot.

CAVALRY

1660	1st Life Guards and 2nd Life Guards	The Cheeses, Tin Bellies, Tins
1661	Royal Horse Guards (The Blues)	The Oxford Blues, The Blue Guards, The Blues
1685	1st (King's) Dragoon Guards	The Trades Union, The K.D.G.'s
1685	2nd Dragoon Guards (Queen's Bays)	Rusty Buckles, The Bays
1685	3rd (Prince of Wales's) Dragoon Guards	The Old Canaries
1685	4th (Royal Irish) Dragoon Guards	The Buttermilks, Mounted Micks
1685	5th (Princess Charlotte of Wales's) Dragoon Guards	The Green Horse, The Green Dragoon Guards, The Old Farmers

1685	6th Dragoon Guards (Carabiniers)	Tichbornes Own, The Carbs
1688	7th (Princess Royal's) Dragoon Guards	Strawboots, The Black Horse
1661	1st (Royal) Dragoons	The Bird Catchers, The Royals
1678	2nd Dragoons (Royal Scots Greys)	Bubbly Jocks, The Bird Catchers, The Greys
1685	3rd (King's Own) Hussars	Lord Adam Gordon's Life Guards, Blands Dragoons
1685	4th (Queen's Own) Hussars	Paget's Irregular Horse
1689	5th (Royal Irish) Lancers	The Daily Advertisers, The Redbreasts, Irish Lancers
1689	6th (Inniskilling) Dragoons	The Old Inniskillings, The Skins
1690	7th (Queen's Own) Hussars	The Old Saucy Seventh, The Lily White Seventh, Young Eyes, Old Straws, Strawboots
1693	8th (King's Royal Irish) Hussars	St.George's, The Cross Belts, The Twenty Fives
1715	9th (Queen's Royal) Lancers	Wynne's Dragoons, The Delhi Spearmow
1715	10th (Prince of Wales's Own Royal) Hussars	Bakers Light Bobs, The Chainy Tenth
1715	11th (Prince Albert's Own) Hussars	The Cherry Pickers, The Cherubims, Lord Cardigan's Bloodhounds
1715	12th (Prince of Wales's Royal) Lancers	The Supple Twelfth
1715	13th Hussars	Ragged Brigade, The Lilywhites, The Evergreens, The Geraniums
1715	14th (King's) Hussars	The Ramnuggar Boys, Pagri Wallahs, The Emperor's Chambermaids
1759	15th (The King's) Hussars	The Fighting Fifteenth, Elliots Tailors
1759	16th (The Queen's) Lancers	Scarlet Lancers
1759	17th (Duke of Cambridge's Own) Lancers	Death or Glory Boys, Skull & Crossbones, Bingham's Dandies
1759	18th (Queen Mary's Own) Hussars	(No nickname!)
1858	19th (Queen Alexandra's Own Royal) Hussars	The Dumpies
1858	20th Hussars	The X's
1858	21st (Empress of India's) Lancers	The Grey Lancers

CORPS

1793	Royal Horse Artillery	The Right of the Line, The Galloping Gunners, The Four Wheeled Hussars
1716	Royal Field Artillery	The Gunners, Field Gunners
1716	Royal Garrison Artillery	Heavy Gunners
1717	Corps of Royal Engineers	The Sappers, The Mud Larks, The Measurers, The Mounted Bricklayers
1888	Royal Army Service Corps	The Moke Train, The Old Wagon Train, Ally Slopers Cavalry
1898	Royal Army Medical Corps	Linseed Lancers, Poultice Wallopers
1877	Corps of Military Police	Red Caps
1878	Army Pay Corps	Quill Drivers, The Ink Slingers
1881	Royal Army Veterinary Corps	The Horse Doctors
1917	Royal Tank Regiment	The Tanks
1920	Royal Corps of Signals	Sigs

INFANTRY

1660	Grenadier Guards 3rd Bn.	Sandbags, Coalheavers, Ole Eyes, Bill Browns
1650	Coldstream Guards	The Coldstreamers
1660	Scots Guards	The Jocks, The Kiddies
1900	Irish Guards	Bob's Own, The Micks
1915	Welsh Guards	The Taffys

Infantry - continued.

1633	The Royal Scots 1st Foot	Pontius Pilate's Bodyguard, The Royals
1661	The Queen's (Royal West Surrey Regt) 2nd	The Sleepy Second, Kirkes Lambs, The Tangerines, The Mutton Lancers
1665	The Buffs (East Kent Regt) 3rd (Raised 1572 The Holland Regt)	The Old Buffs, The Nut Crackers, The Resurrectionists
1680	The King's Own Royal Regt 4th (Raised 2nd Tangier Regt)	Barrell's Blues, The Lions
1674	The Royal Northumberland 5th Fusiliers (Raised 1673 The Holland Regt)	The Fighting Fifth, Old and Bold, The Shiners
1688	The Royal Warwickshire 6th Regt. (Raised 1673 The Holland Regt)	The Warwickshire Lads, Saucy Sixth, Guise's Geese
1685	The Royal Fusiliers 7th	The Elegant Extracts
1685	The Kings Regt. 8th	The Leather Hats
1685	The Royal Norfolk Regt. 9th	The Fighting Ninth, The Holy Boys
1685	The Royal Lincolnshire 10th Regt.	The Poachers, The Springers
1685	The Devonshire Regt. 11th	The Bloody Eleventh (Casualties at Salamanca)
1685	The Suffolk Regt. 12th	The Old Dozen
1685	The Somerset Light Infantry 13th	The Illustrious Garrison (Jelalabad)
1685	The West Yorkshire Regt. 14th	The Old and Bold, Calverts Entire
1685	The East Yorkshire Regt. 15th	The Snappers, Poona Guards
1688	The Bedfordshire and 16th Hertfordshire Regt.	The Old Bucks, The Peace Makers, The Bloodless Lambs
1688	The Royal Leicestershire 17th Regt.	Bengal Tigers, Lily Whites
1683	The Royal Irish Regt. 18th	The Namurs, Paddy's Blackguards
1688	The Green Howards 19th	Howards Greens, Howards Garbage
1688	The Lancashire Fusiliers 20th	The Minden Boys, Kingsley's Stand, The Two Tens
1678	Royal Scots Fusiliers 21st	Fusil Jocks, Earl of Mars Grey Breeks
1689	The Cheshire Regt. 22nd	The Two Twos, 1795 The Red Knights, The Lightning Conductors, Meeanee Boys
1689	Royal Welch Fusiliers 23rd	The Nanny Goats, The Royal Goats
1689	The South Wales Borderers 24th	Howards Greens (1717-1737)
1689	The Kings Own Scottish Borderers 25th	Kosbys, The Botherers, Kokky-Olly-Birds
1689 1794	The Cameronians 26th (Scottish Rifles) 90th	Sir Thomas Graham's, The Perthshire Grey Breeks
1689 1861	Royal Inniskilling Fusiliers 27th " " " 108th	The Skins The Lumps
1694-28 1758	The Gloucestershire Regt. 28th " " " 61st	Fore and Aft, Back Numbers, The Old Braggs, The Slashers
1694 1701	The Worcestershire Regt. 29th " " " 36th	The Firms, Old and Bold, The Star of the Line, The Saucy Greens, The Brummagon Guards, The Eversworded, Vein Openers
1702 1755	The East Lancashire Regt. 30th " " " 59th	30 Triple X's, The Three Tens, 59 Lily Whites
1702 1758	The East Surrey Regt. 31st " " " 70th	The Young Buffs Glasgow Greys
1702 1741	The Duke of Cornwalls Light Infantry 32nd " " " " 46th	The Surprisers, Red Feathers – 1777, The Docs, Murray's Bucks, Lacedaemonians
1702 1787	The Duke of Wellington's Regt. 33rd " " " " 76th	The Dukes, The Havercake Lads – 1814, The Old Immortals, The Hindoostan Regt., The Old Seven and Sixpennies

1702	The Border Regt. 34th	The Two Fives, The Cattle Reeves
1755	” ” 55th	
1701	The Royal Sussex Regt. 35th	The Orange Lilies (1701 - William III)
1854	” ” ” 107th	The Belfast Regt.
1702	The Royal Hampshire Regt. 37th	The Hampshire Tigers
1758	” ” ” 67th	
1702	The South Staffordshire Regt. 38th	Pump and Tortoise
1793	” ” ” ” 80th	The Staffordshire Knots
1702	The Dorset Regt. 39th	Sankey's Horse (1707), Green Linnets (mounted on mules)
1755	” ” 54th	The Flamers (1781)
1717	The South Lancashire Regt. Prince of Wales Volunteers 40th	The Excellers (XL), The Fighting Fortierth
1793	” ” 82nd	
1719	The Welch Regt. 41st	The Fogies, 1st Invalids (1719)
1758	” ” 69th	Ups and Downs, The Old Agamemmnons (Lord Nelson) Wardours Horse
1730	The Black Watch 42nd	Forty-twas, The Watch
1780	” ” 73rd	
1741	The Oxfordshire & Buckinghamshire Light Infantry 43rd	The Light Bobs
1755	” ” 52nd	
1741	The Essex Regt. 44th	The Little Fighting Fours, The Two Fours,
1755	” ” 56th	The Pompadours (1755), Saucy Pompeys
1741	The Sherwood Foresters 45th	The Old Stubborns (Talavera)
1832	” ” ” 95th	
1741	The Loyal Regt. 47th	The Cauliflowers (facings), The Lancashire Lads, Wolfes Own (Quebec)
1793	” ” 81st	Loyal Lincoln Volunteers
1741	The Northamptonshire Regt. 48th	The Steel Backs (floggings)
1755	” ” ” 58th	Black Cuffs
1743	The Royal Berkshire Regt. 49th	The Farmers Boys
1758	” ” ” 66th	The Biscuit Boys (depot at Reading)
1755	The Queen's Own Royal West Kent Regt. 50th	The Dirty Half Hundred (black facings), The Devils Royals 1807, Blind Half Hundred
1824	” 97th	The Celestials (sky blue facings)
1755	The King's Own Yorkshire Light Infantry 51st	The Kolis
1839	” 105th	
1755	The King's Shropshire Light Infantry 53rd	The Brickdusts (facings), Old Five and Threepennies
1793	” ” ” ” 85th	Elegant Extracts
1755	The Middlesex Regt. 57th	Die Hards (Albuhera), Steel Backs
1787	” ” 77th	The Pothooks, Two sevens
1755	The Kings Royal Rifle Corps 60th	60th, Green Jackets, The Jaegers
1758	The Wiltshire Regt. 62nd	The Splashers (buttons), The Moon Rakers,
1824	” ” 99th	The Springers
1758	The Manchester Regt. 63rd	Blood Suckers
1824	” ” 96th	Bendovers
1758	The North Staffordshire Regt. 64th	The Black Knots
1824	” ” ” 98th	
1758	The York and Lancaster Regt. 65th	Royal Tigers
1793	” ” ” 84th	Twin Roses
1758	The Durham Light Infantry 68th	The Faithful Durhams
1839	” ” ” 106th	
1777	The Highland Light Infantry 71st	The Glesca Keelies
1787	” ” ” 74th	The Assaye Regt.
1777	Seaforth Highlanders 72nd	The Macraes
1793	” ” 78th	The Kings Men

1787	The Gordon Highlanders 75th	The Gay Gordons
1794	" " 92nd	
1793	The Queens Own Cameron Highlanders 79th (2nd Bn 1897)	The Camerons
1793	The Royal Ulster Rifles (late Royal Irish Rifles) – 83rd	Fitch's Grenadiers
1793	– 86th	Irish Giants
1793	Royal Irish Fusiliers 87th	The Old Fogs, Faugh-a-Ballagh Boys, The Eagle Takers
1793	" " 89th	(Barrosa), The Rollickers, Blayney's Bloodhounds
1793	The Connaught Rangers 88th	The Fagon Bealachs, The Devils Own (Picton)
1823	" " 94th	The Garvies
1794	The Argyll and Sutherland Highlanders – 91st	The Rorys
1800	– 93rd	The Thin Red Line
1858	The Leinster Regt. 100th (Canada)	The Crusaders, The Centipedes, The Beavers, The Colonials
1853	" " 109th (India)	The Poonah Pets, The Steel Heads
1756	The Royal Munster Fusiliers 101st (India)	The Dirty Shirts
1839	" " 104th (Bengal)	Delhi 1857
1748	The Royal Dublin Fusiliers 102nd (India)	The Blue Caps
1661	" " 103rd (Bombay)	The Old Toughs
1800	The Rifle Brigade	The Sweeps, The Green Jackets
1940	The Parachute Regt.	The Red Devils

xiv

The illustrations and their stories

by Lt-Col Frank Wilson

Page 29

The President of the Mess Committee of a Royal Regiment inspects the seating plan and table before a regimental Guest Night.

Right to left: Mess President, usually a senior Major; the mess sergeant with seating plan; the mess corporal finds the famous Burmese Bowl is not up to standard; a mess waiter, who is also 'silver cleaner' taking away a band programme, also not up to standard; the cook (civilian locally employed) attempting to attract the attention of the mess sergeant to inform him that the sole ordered from the local fishmonger has not yet arrived.

Note: The colours on other nights would be 'cased'.

Page 18

The Senior Subaltern welcomes the new addition to the Officers' Mess: circa 1890 "Knew yer father, hope you are as good an officer as he was".

The new officer's luggage consists of a top hat, in case, his father's gun, helmet case, sword (in case), valise for manoeuvres (with blankets & sheets), uniform metal box, gladstone bag.

Gentlemen never travelled without a rug – foot warmers could be hired.

Trunk with plain clothes, tail coat, morning coat, etc. is still in guards van and will go on with tin trunk by G.S. wagon (horse drawn) later.

Notes: A – Subalterns could do up to 16 years before they became a Captain. They loved their regiment and had no wish to transfer to another regiment where promotion was faster. B – Most regiments had 'dog carts' (horse drawn transport) and also married officers. There were no motor cars, if married officers went out to dinner, then a cab from local livery stables took them.

Page 14

SCENE – A Subaltern's Quarter in Officers' Mess, circa 1910.

Left to right: Great coat; cartridge bag and father's gun; field glasses (used mostly when at races); helmet; group photo; first fox brush; fox mask; crop, pipes and hunting print; oil lamp; hunting map*; mug for morning tea; blanket and table, G.S.; training manuals (un-read); hunting boots, spurs, tennis racket, fishing rod and sword.

* Hunting map with circles in miles to show distance of 'meets' from barracks.

On mantlepiece – 'Mum', 'Sis', tobacco tin, invitations, cup for point to point, 'Dad'.

Below: kettle for hot water bottle; brigg's umbrella; mess jacket and waistcoat; blankets; G.S. sheets – officer's for use of; pots, chamber – officer's for use of. Mess kit laid out by soldier servant (not batman in those days) now on coal fatigue. Window left open in order to draw up fire – rain pours in.

Note: Barracks probably built about 1810.

Page 53

The Guest of Honour

Governor of the Province, ruler of millions of Indians, has 'Angels on Horseback' poured down his front by a new mess Khitmagar.

Left to right (seated):

The Private Secretary *(thinks)*
"Now he will be in a bad temper at the meeting tomorrow with those congress chaps I had so carefully arranged."

The Second in Command
'Angels on Horse Back' are his favourite dish, his portion is now down His Excellency's waistcoat. Next lot will be cold. At his age, as he tells the subalterns, there is nothing to enjoy except the pleasures of the table.

His Excellency *(thinks)*
"What will dear Mabel think when she sees the new white waistcoat with gold buttons she put in my stocking at Christmas".

The Colonel *(thinks)*
"This will be the joke of the cantonment Gymkhana Club."

The ADC *(does not think too much)*
On His Excellency's staff because he is a charming chap and has a handicap of 6 at polo – most useful to Governor's team. He is Indian Cavalry and naturally has a great sense of humour, he has seen it happen before, not in his own regiment, of course.

Background –

The Abdar
Who knew something like this might happen, but not to the 'Lat sahib', arrives with bowl and sponge from the mess sergeant's bunk to wash down His Excellency.

The Governor's Private Servant
Who always stands behind him at meals, his chief duty being to pull his chair back when he wants to sit down and then push it forward. He is not surprised. He also has seen this happen before and has a napkin ready.

The Khitmagar
His first Guest Night. Far better to go back to his village and drive a bullock cart.

The Mess Sergeant
Hissed abuse almost fails him as he points out to the 'accused' that the Governor and Commanding Officer have already got their 'Angels on Horseback'.

Page 33

01.35 hrs. A Regimental Guest Night for the Brigadier and his staff, the Master of Foxhounds and a rich landowner. (Between the two great wars. The band have gone home.)

Right to left:
A. The Mess waiter – 'Basher' Barker (or 'Basher boy') employed in the Officers' Mess so that he will get better food to strengthen him for the finals of the army middleweight championship and so gain credit for the regiment. He has enjoyed the evening enormously, at the expense of a few nips on the regimental guest's fund. He does not seek promotion, he does not fear man nor beast, nor generals, or such like. All he wants, in his simple way, is to get the Brigadier to accept his brandy before he, 'Basher', goes to his 'bunk' for a deep sleep and 'roadwork' under a P.T. Sgt. at 07.00.

B. The Master of Foxhounds *(thinks)*
"A nice lot of chaps – must reduce the hunt subscription for these keen young boys from the camp."

C. The Brigadier
Known to the troops as 'Stop me and buy one', now mellowed, is telling his favourite story about a 'grass widow' on a houseboat in Kashmir.

D. The Colonel
Doubled up with laughter by the Brigadier's story which he has heard twice before. It is important that the Brigadier enjoys his visit to the regiment as the Colonel longs for promotion, having three sons at a public school.

E. The Second in Command
A gallant officer – rather 'fond of the bottle' but has no axe to grind. Spent all his service in the regiment. Is a bachelor and has little to look forward to except a bed-sitter in the Cromwell Road, long chats about old days in the bar of his club, and cheap meals in the 'new fangled' snack bar. Known kindly by the troops as 'Old Daddy Boy'.

F. The Brigade Major
A keen polo and billiards player – can concentrate (hence P.S.C.) but has heard his Brigadier's story 47 times. Too ambitious to show his feelings.

G. The Local 'Box Wallah' or industrialist who has the best shoot in the county to which officers hope to be invited. He is being 'made much of' by a captain who is an excellent shot but too poor to afford a gun in a syndicate.

H. In the background – the 'warts' or subalterns. They are unable to afford a drink or leave till all the guests go. Other 'warts' are asleep in the ante-room or playing simple rough games. Later they will wake up and play 'billiard fives', no doubt.

The more serious of the elderly are playing bridge in the silence of the bridge room.

A few crafty subalterns have crept away – probably to be dragged out of bed later by those who have stuck out the long night.

Page 45

Behind the Officers' Mess bungalow – Pay day for the Mess servants. 1920 – 1939.

Left to right:

1. The Mess Corporal with rupees and annas. He loves a bit of a 'tamasha' and hopes there will be a fight.

2. The Mess Sergeant, not long out from the Home Battalion, is bewildered. His Urdu is limited to a few phrases, as was Lady Mabel's in the well known poem –

The Lady Mabel's words were few.
They could not well be fewer.
They were 'Idhar aou'[1]
'Jaldi Jad';[2] and 'Khibidar you souer'.[3]

[1] – Come here; [2] – Go quickly; [3] – Take care you pig.

With the Mess Sergeant is his dog 'Heinz' (57 different varieties) passed on to him by his predecessor – always the terror of the compound – he will bite anybody from Generals to 'untouchables'. He knows no 'caste'.

Dogs in North India always wore miniature saddle cloths at dusk – as the temperature fell fast. These were made very carefully by the Durzee (tailor) and embellished with regimental colours and crest. Note: Nobody had the courage to take Heinz's off.

3. The Head Khitmagar, Abdar, that is head of all the mess servants. He is the interpreter having served the two battalions of the regiment in India all his life – he knows all.

Mohd Ali Khan knows well that Lab Chand, the cook, is most comfortably off and has bought a teashop in the suddar bazaar where he sells unwanted food from the officers' mess, with tea, to soldiers of the regiment.

4. The No.2 Khitmagar, or waiter, known as 'The Camel' who hopes that Mohd Ali Khan will go too far and he will get his job.

5. The Khansamah, or cook. The fattest man in cantonments and is 'on contract' for messing.

"I am starving fellow with wife and seven little children, all most underfed. Selfs large size due to stomach trouble due to long service under benign government on North West Frontier – Assistant Surgeon will confirm – how can I live if some officers eat six eggs for breakfast – Please to increase pay before entire family perish from starvation."

6. Masalchi or washer up – cousin of cook. Only complaint is he needs new drying up rags and pots and pans.

7. Bhisti – or water carrier – known as 'Gunga Din – complains that he needs a new mussock or goat's skin bag, used to get water from the Mess well, also that kerosene oil cans are leaking (used for water).

8. Mess 'Shikari' of officers' gamekeeper with present of cotton teal for sergeant's mem sahiba's dinner (bird shot by officer, but mislaid).

9. Head 'Mali' or gardener with rare plant (from mess garden) for sergeant sahib's quarter.

10. In background, the Dhobi, or washerman. He has forgotten the time as he has no watch and has been hard at work beating napkins, tablecloths and officers' stiff shirts on a boulder in a muddy 'nullah' two miles away. He has left his donkey outside the mess as it eats the few flowers in the garden.

11. In the foreground is the Chowkidar or night watchman, Till the officers went to bed he smoked his hookah, coughed and shouted 'Khabidar' (take care). He then slept, overcome by his long vigil – he has now dropped off again. But he wants his pay and needs oil for his hurricane butti (lamp).

He is still dressed in his working clothes, a greatcoat, U.S., and blanket O.R.'s for use of, also U.S. (unserviceable). Well in the background is the mess sweeper, an 'untouchable', he may not come near the higher castes. Dressed in cast off clothing he carries the badges of his trade, a brush and commode basket. He has no complaints, only that he may carry on working – known by all servants, sarcastically, as 'Jemadar' (an Indian officer equal to a 2/lieut.).

12. In the background the new addition to the mess, just arrived. After a route march all he wants is a large shandy to wash the dust of India out of his system. He thinks, judging by the noise, it must be one of those communal riots of which his uncle in the Indian Army had told him.

Notes: Undeterred by the noise, and unnoticed, wild bees are swarming on the verandah roof.

The mess cookhouse is about 200 yards away to the rear of the mess. Officers' mess cook houses had the reputation of being the most unsavoury in the lines. Food was carried to a hot box where it was supposed to be kept warm by a small charcoal fire. The hot box can be seen on the verandah behind the 'Abdar'. The sergeant's and corporal's tea is on top.

Windows of the scullery, silver room, wine store, etc. had wire gauze over them to keep the flies out. This was usually broken, so that flies could get in, but not out.

On an inspection fly traps in old kerosene oil tins were put anywhere where they could be seen and the sweeper would laboriously fill them with flies he had caught.

Page 19

An Officers' Mess Ante-Room at a School of Instruction c.1895

The officers, after a late lecture, are drinking sherry before dinner. 'Cocktails' or 'Rotgut' as they were called, were not drunk. Gin was considered 'a drink of the lower classes, charladies and persons of that sort'. (Remark by a General of Foot Guards on being offered a pink gin in a mess in India just after the first World War.)

Two embarrassed young officers are creeping out of the mess to their 'growler' in order to catch a fast train to London, so that they may attend Lady Mabel Marshmallow's debutante ball in Eaton Square, for her fourth daughter. (Note the silk hats.)

The officer on the right is a cavalry officer, as can be seen by the white 'swagger mark' across his forehead. This was due to officers of mounted arms wearing the 'pill box' hat on one side of the head. It also left a white chin strap mark. They could be picked out when in tails at the theatre or at a dance, and they enjoyed it.

If very rich the young officers, with others, would order a 'special train' of one coach and a steam engine, rugs and footwarmers supplied, to take them back. They would be seen off by the assistant station master in a top hat.

If they were poor, as was usually the case, they travelled from the dance to the station, in a hansom cab and caught the early 'milk train'.

Arriving at dawn, they walked back to their mess in time for breakfast and an early lecture through which they dozed or slept.

Enjoyable though mess life was, there was one drawback, there were no secrets. Every member of a mess knew the private affairs of all the other members.

The mess dresses shown are from left to right: 'Heavy Cavalry', Royal Field Artillery, Foot Guards, Rifles, a Scots Regiment, and the Royal Engineers.

In the foreground is an elderly instructor on the staff. An expert on ballistics. Nobody asks him to dances now, but he remembers the good old days of free dinners, dances and suppers, followed perhaps by a visit to 'Romanos' where, till dawn, he would sip champagne with Phil May and some lovely girls from the front row of the chorus who, as the evening wore on, would even show their ankles.

Note: Growler – a cab.

Page 44

An Interruption to the Ritual of the Tying on of the Cummerbund during the hot weather in N.India.

Officers in India wore white mess jackets and cummerbunds in the hot weather. Some regiments wore white drill overalls, others trews or overalls of blue serge.

The officer revolved gracefully like a ballet dancer. His bearer, large safety pin clenched in remaining teeth, then secured it behind his 'master'.

In the Indian Cavalry the cummerbund was usually the full dress lungi (turban) of a Sowar (trooper) of his regiment. In the British Army it was often shorter, and of the facings of the regiment.

The scene shows an officer of the Indian Cavalry who looks disturbed although looking forward to his iced soup in the garden of the mess, even if it contains the wings left by flying ants.

The Regimental Orderly arrives with a secret message from the Adjutant Sahib Bahadur. It states that 'C' Sqn. will march at 21.00 hrs. down the Grand Trunk Road to Fort Govindgarh, outside the walls of Amritsar for communal riot duty. Details to follow.

The Orderly knows all this as the office Dafadar, Mahbub Ali, comes from the same village north of the Jhelum River, and married his uncle's daughter.

He assures the officer who has just taken over command of 'C' Sqn. that Risaldar Sultan Khan Sahib, the senior Indian officer in the squadron, has everything under control. He is also a relation and knew, in his wisdom, that this would happen. Horses will be watered and fed, feed bags filled and Chaupattis are now being cooked for haversack rations. Water bottles have been filled, arms and ammunition will be issued from the rifle kote at a time suggested by the Captain Sahib Habadur.

The orderly has also taken the liberty to call at the officers' mess on the way to inform the mess Dafadar that the captain sahib would need sandwiches and two bottles of beer 'sharab' – to be iced, till called for.

This is not news to Sowar Sher Khan, the captain's orderly. He knew what was brewing from chat he heard in the bazaar that afternoon.

The captain's charger is ready and saddle prepared. He himself is equipped to happily destroy either side and delighted to have a change from schooling the captain sahib's polo ponies.

The captain's bearer, Gulab Mohd, a Pathan who looked after the 'chota sahib's' father, a really splendid man, who became a 'burra sahib', is not pleased. He will have to lay out his marching order, take down the mosquito net, pack his bedding roll, keep the creditors at bay and see that his 'Shaitan' (devil) of a dog is alive on his return.

A British regiment of foot is also going to march, but not so fast.

No Officers of the army were so well cared for by their men as those of the Indian Army. For 'sowars' and 'sepoys' reasoned that if their officers were 'kush' (happy) then they would not be cross with their soldiers. They were also trusted and were fond of their officers, for in time of trouble they were 'man-bap' (mother and father) to them.

Notes:

1. The only telephones in a regiment were usually to the colonel's bungalow, the second in command, the adjutant and the mess.

2. Invitations to tea and tennis 'The Dancants' at the club, lunch, dinner or cocktail parties, were sent by 'chit', by hand of the Mali, Masalchi or some unimportant servant.

3. This was a good practice, as the recipient in the cold weather had time to make out an excuse, i.e. not to meet 'some jolly girls', who he knew were plain and dull, and had come out from England with the 'fishing fleet'. The one excuse that never worked was 'I am afraid I am orderly officer'. The mem-sahibas knew who was orderly officer.

4. Dogs in India always had their miniature charpoys (beds) to keep the creeping insects away from them. They lay stretched out to keep their stomachs as cool as possible.

A busy stream of ants goes back and forth in military procession to collect bits of cake dropped by the captain sahib, as he had his tea before a few chukkers of slow polo with other officers and their orderlies.

The rains have broken and outside, and in, it is hot and humid. Flying bugs are everywhere, especially round the lights.

5. The room in a bungalow, shared with other officers, is large and high. From the ceiling there would be an electric 'punkah' or fan.

6. All furniture, charpoy (or Indian bed), dhurrie (carpet), curtains, tables, table fan, etc. are hired from the 'suddar bazaar', the only extras the officer owns is the dog's bed, a rug he bought from a Kabuli carpet wallah, his pictures, some heads of wild animals he shot, tushes of a giant pig he got as 'first spear' out pig sticking, his bedding and clothes — and probably a few debts to friendly money lenders in the bazaar who lent rupees at 3% a month, or 4% if the sahib was in England on leave.

Page 49

An Officer's bathroom, or 'Ghusal Khana', in India.

On the orders of the bearer, the 'Bhisti' is bringing in two old kerosene oil tins containing cold and hot water for the sahib's bath.

He treads carefully, and wisely, as that deadly snake, the Krait, enjoys the cool of a bathroom.

Notes:
When a bath was done with, the 'Bhisti' tipped it up and the water ran out into the garden, through a hole in the wall, the entrance for the snake and various types of insects.

The bath is surrounded by a low mud and brick wall, in the corner is a large porous 'chatti' or pot, that holds cold water.

No sponges were used as scorpions dwelt in them, tin mugs were used for the sahib to sluice himself with and to get cold water, if needed.

The hum of mosquitoes was loud in a bath room in the hot weather.

In the background is the 'thunder box' or commode. This was dealt with by the sweeper. On the wall a cardboard box of bronco toilet paper.

Young subalterns going East in a troop ship were warned by old pukka sahibs or 'Koi-Hais' of the Krait and how it lodged on top of the door and fell on the officer and bit. That, they sighed, would be the end of a promising future in a few minutes. They never explained how the Krait got up there.

Page 43

Diagram of an Officers' Mess in India.

The walls of old messes in India were made of bricks and dried mud, about 18" thick. Later new messes of red brick were built but were not nearly so cool.

There was always a vast porch in front of the mess under which, in the old days, a horse, 'syce' and 'tum-tum' could wait in the shade.

The front or sides of the bungalow had a verandah decorated by the 'malis' with flower pots. The roof of the bungalow was supported by pillars or curved arches. 'Chicks' or screen blinds of finely split bamboo, laced by string, and usually painted green outside, were let down to keep out the glare of the sun.

The rooms were high and the ceilings supported by wooden beams riddled by white ants.

The floors were of dried mud and covered by 'chatai', a closely woven rush matting above which there was a dusty 'dhurrie' or carpet.

Behind the dining room there was a small verandah. During a meal it was the scene of much activity and squabbling as servants attempted to get at the 'hot boxes' to procure luke warm food for their masters.

The kitchen was about 100 yards to the rear of the mess and hot food had to be carried to the 'hot boxes', kept warm by charcoal. 'Kite hawks' did their best to swoop down to get at it. In a British regiment the mess corporal usually slept in dignified isolation in the heavily barred silver room.

The servants' quarters were well to the rear, made of dried mud with 'purdah' walls all round. In the small dark rooms the servants, their families and a few poorer relations lived in noisy happiness.

The well was worked by two bullocks and usually of the 'Persian wheel' type. The bullocks went round and round, blinded by blinkers in case they got dizzy, drawing up small earthenware pots of water. The water was diverted into narrow mud channels to water the lawns and kitchen garden.

As the Persian wheel went round it made a pleasant moaning sound which will be remembered by all who served in the East and induced sleep to all including the bullocks and their driver with his sharp prong used to urge them on.

Pages 25, 26

St.David's Day is celebrated in peace or war by the Royal Welch Fusiliers wherever they may be – Flanders 1915.

After a picture by S.T.Dadd in possession of the regiment.

The Regimental ceremonial of eating the leek has remained almost unchanged over the years. It was described in detail in the records as having taken place at Boston, Virginia, in 1775, and referred to then as an ancient custom. After dinner the Goat is led around the table followed by a drummer, fifers, the drum major carrying a silver salver on which are leeks, and the mess sergeant with a loving cup charged with champagne. They halt by the most recently joined Subaltern who stands on his chair and placing his right foot on the table is offered a leek. Whilst he eats the leek the drummer plays a continuous roll until it is completely consumed. The mess sergeant then hands him the loving cup for a toast to St.David. All those present who have not eaten a leek with the Regiment (including the guests), are expected to eat one before the party withdraws.

The five black ribbons which the Regiment wear on their collars are a link with the days when soldiers wore powdered and greased pigtails. In order to protect the red tunics the pigtail was enclosed in a 'queue bag'. When the queue was abolished in 1808 the Regiment was serving in Nova Scotia and as communications were slow in those days it is probable that the Regiment was the last to wear it. When, however, they finally dispensed with the queue the ribbons with which it was secured were retained and these were known as the 'Flash'.

Page 2

Lunch at Barwa

This sketch comes from 'Sketches on Service' by Major E.A.P. Hobday, RA on a frontier campaign just at the turn of the century. He describes it thus......

At Barwa we spent the heat of the day in a lovely chenar grove by a spring just below the fort; and here we were entertained by the Khan, who had charpoys covered with carpets and rugs set for us, and soon strings of retainers appeared with bowls and baskets containing all manner of refreshments – rice, pillaus, curry, stewed kids, baked fowls, Afghan bread (very different from the ordinary chaupattis), grapes, apples, and all kinds of good things. We graciously accepted a share of these and the remainder were handed over to the escort of whom the Mussalmans fared sumptuously.

The members of the party besides myself on this occasion were: – Major Deane, Political Officer; Major Beatson; Major Delamain; Lt.Waterfield, 11th Bengal Lancers; Lord Fincastle, 16th Lancers; and Lt.Churchill, 4th Hussars (2nd from right).

Page 62

Hunt the 'Po' (The Kill)

This sobering and exciting pastime was carried out by the subalterns of the Royal Scots Fusiliers after Guest Nights.

(The anti-blood sports associates could not have 'faulted it', in spite of the fact that the officers had indulged in some sort of blood sports that morning.)

An officer who had a reputation for 'pushing a pretty pedal' mounted a bicycle (probably the mess corporal's) with a tin chamber pot tied by cord to the saddle.

He set off from the mess and was given a few minutes' start. The pack then followed the 'scent' by the tinkling sound of the 'pots, chamber', hitting the ground.

'The fox' usually went to earth in some other mess which he knew well.

Page 13

Unforgivable mistakes made by a 2nd Lieutenant on his first Guest Night in the Mess.

1. He is sitting on the 'bum warmer' reserved for senior officers; junior subalterns hovered in that part of the ante-room furthest from the fire.
2. He is talking.
3. He is smoking a pipe before dinner.
4. He is drinking a pint of beer (or wallop).
5. It is suspected his tie is 'made up'.
6. He wears an eye glass.
7. He does not hunt, shoot, play polo or any other games except 'smack sparrow' (badminton).
8. He likes singing and amateur theatricals and moving pictures.
9. He was heard playing the accordion.
10. His hair is too long.
11. His conversation is about 'the dancants' and 'flappers'.
12. To the second in command he says "take a pew old boy and have another on me".

He won't last long at this rate.

The waiter who has been in the mess for many years and knows his 'etty-quetty' better than most, creeps out before the coming eruption (or explosion) shatters the pictures in the ante-room. (Even the clock, presented by a crowned head, has stopped.)

Page 74

A 'Shooter Sowar' takes out beer and 'sandbeefs' to members of an Indian Cavalry Mess on a duck shoot.

'Does not he alone know the hundredth attribute of God which has been lost by man and is not his supercilious air due to the possession of this secret.'

The camel was unloved by man and horse, but the former should be grateful to him.

There were usually eight 'Sowari camels', that is camels that could be ridden, to a regiment of Indian Cavalry.

They were known as 'Shooter Sowars', from the Persian 'Shutur-Sawar', the rider of a swift camel.

These camels had bonnets to keep the flies off, saddle cloths and round their necks a bracelet of bells. These made a tinkling sound as they trotted down the tan paths of cantonments, on their way to deliver or collect letters from the various headquarters and regiments. Horses on this duty were apt to 'run up light' due to waiting outside the various offices, as some officer fumbled through his out tray for a letter blown into a corner by 'the Punkah'.

They were most useful on manoeuvres and the line of march. Transport was by bullock carts which moved at 2 mph or baggage camels which were as slow. The Sowari camels, to the envy of the British Cavalry, were able at the long halt, after three hours marching up a dusty road, to deliver beer and sandwiches (always known by Indians as 'sandbeefs') to the officers. The Sowars had 'chaupatties' in their haversacks and their water bottles.

They also were able to take out food and refreshment to officers and their guests when pig sticking or shooting. The picture shows a 'Sowari camel' arriving late, due to the rider having no watch, and stopping in a village to have a pull at a 'hubble bubble' (or hookah), and listen to a choice bit of scandal about the local police inspector.

The mess Khitmagar shouts "son of an owl, you are late, the officer sahibs are dying of thirst and hunger and very angry. They have been here standing in water since before dawn. They are about to drop with fatigue, and I, Ghulam Nawaz Khan, have been unable to find a place to eat in this Hindu district".

Page 55

Drama in an African Mess

In the days of our Empire there were officers' messes all over the world in which, however small, the members kept up the old traditions. Some were large and well built, others were in forts, many in Africa were of mud and thatch. Officers of the British Army could be seconded to regiments in Africa where they got more pay and a better chance of malaria.

The majority of officers went in order to pay off their debts, others in order to save up to marry some lovely girl they had proposed to at The Berkeley, a few for adventure and a change of scene, while one or two who did not see eye to eye with their Colonels were 'sent'.

The officers had to pass colloquial exams in the language of the colony they served in but the mess servants usually spoke 'pigeon English'.

The scene is a mess on the West coast of Africa. The mess secretary is sipping his 'coaster' before dinner, when No.1 boy rushes into the ante-room brandishing a carving knife and shouting 'Buttah he no live for cupbardh – cookoo he say I be thief man – he always plenty humbug me – he no fit for live – I go chop him'.

These dramatic incidents were treated with the imperial calm they deserved. If a fight did occur both parties were separated and told to lie face downwards. The sergeant major was sent for (usually a tall Hausa from the edge of the Sahara), and he administered 'Five for Arse' with his leather belt.

This form of summary punishment, not in the manner of military law or King's Regulations, was also dealt out to young soldiers when they squabbled. They preferred it to going before their company commander as it did not appear on their crime sheets. The spectacle was also much enjoyed by their comrades.

Note: A 'coaster' was a pink gin with soda water. Forty years ago the only place where you asked for a 'coaster' and got what you wanted was from 'Blossom' in the 'Premier Lounge' off Piccadilly.

There 'traders' and officers from the west coast sat talking wistfully of 'ground nut stew' and 'palm oil chop' and how much quinine they still had to take.

Page 60

The Orderly Officer's Vigil Scene – Officers' Mess Ante-Room – Midnight – India – A Scots Regiment

The orderly officer is worn out after a long day. Before breakfast he smelt the meat at the ration stand. This made him feel rather sick, due to a splendid Guest Night with the Queen's Bays the night before.

He then visited all the cook houses and 5 companies' breakfasts. All were as usual 'bergue' (porridge) and fried eggs swimming in 'ghi' (or native lard). Fusilier 34167 'Smudger Smith', a well known producer of 'frivolous complaints' said the eggs 'tasted ever so queer'. The orderly officer forced himself to taste the revolting dish, covered in pepper and salt by 'Smudger'. This nearly finished him. He retired to the mess and managed to force down some 'rumble tumble' (scrambled eggs).

At 09 00 hrs. with the Adjutant, he attended guard mounting outside the orderly room. The Adjutant, a keen officer, who did not try to understand how our hero was suffering, told him to pick out the smartest man to be 'stick orderly'. This coveted post meant that the Fusilier, wearing a pipe clayed leather bag, with a large grenade on it, bicycled slowly round cantonments with messages, stopping here and there for cups of tea.

This he found impossible as all were equally well turned out; he even looked to see if the nails on the soles of their boots were polished. Leaning down brought on another turn, so he chose a man from his own platoon.

He then indulged in sectional rushes with his platoon in the vicinity of the stinking nullah which lay outside the lines. This was followed by 45 minutes' semaphore, which he never mastered, but would be useful perhaps in the next station at the summit of the Khyber Pass.

He was given a mug of hot 'char' by the kindly company clerk, in the clerks' 'bunk' which was hung with bazaar postcards of himself, in various full dress uniforms provided by the photographer. 'Nobby' Clarke liked the one of himself in an ancient Royal Horse Artillery jacket, as did Jeanie, his girl friend in Ayr, who he had not seen for five years. "Roll on ma boat", he moaned.

This was followed by a visit to education on the bungalow verandah under the educated Corporal Dearlove. The note-book's 'for sums' stuck to the tables, still wet from scrubbing by mess orderlies.

At 12 45 cookhouses and fly traps were peered at and, after the orderly sergeant had banged his swagger stick on the tables, the bully beef rissoles in all companies were viewed.

After a late tiffin in the mess, he went to his bungalow and changed into polo kit, as he had got a 'hockey player' to answer for him till 16 30 hrs. Four chukkers of polo, on his country bred troop horse B43, that carried a machine gun usually and was kindly lent to the regiment by the Indian Cavalry.

Having changed back into uniform he bicycled to the lines at retreat to mount mule stable guards, the inlying piquets and a guard on the RHA lines. (The battery were firing their guns off miles and miles away.)

After dinner in the mess, at 22 00 hrs. he attended staff parade, and then 'biked' back to the mess to report to the senior officer who was playing bridge, and losing. "Battalion reported present, sir, with exception of one old and two new absentees." The senior officer did not even look up, just grunted.

The captain of the week has unfortunately decided to go round the guards at 23 00 hrs. so the orderly officer must wait till midnight before he can visit them. He settles down in an empty ante-room with the civil and military Gazette and a four week old 'Tatler'.

The bewitching hour of midnight has arrived, even the chowkidar (night watchman) has been asleep for 2 hours. The mess corporal longs to lock up the mess, go to his bunk and take off his dicky.

Alert, as a result of the Khitmagar's respectful prodding, the orderly officer gets on his bicycle and sets off under a starlit sky which he does not appreciate, off down the dusty road past the vast company bungalows, white in the moonlight. The jackals howl round the cookhouses their theme – Dead Hindoo – where – where – where?

He approaches the quarter guard with care. There is a rattle of chains as the sentry brings down his musket loaded with buckshot to the 'on guard' and shouts "who goes there". After a loud answer of "visiting rounds" there is a cry "Turrn oot ther guard" and the guard who had just dropped off after 'grand rounds' sway out, breathing heavily smelling of sleepy fug and blankets and form up.

The orderly officer then bicycles two miles to inspect his battalion's guard on the empty RHA lines.

After leaving those lines he gets a 'jubiduster' or prickly Indian thorn in his front tyre. He walks back to his bungalow (the bicycle is, of course, hired, as is his furniture).

He goes to bed and sleepily thinks "all for ten shillings a day – but father said I would make no money, but would have lasting friends and the tradition of a splendid regiment 250 years old behind me, all for half a day's work".

He sinks into a deep sleep to be woken up at 06 00 hrs. by his bearer with a cup of tea, covered in oily circles of buffalo milk, before taking his company for 'a run' at 07 00 hrs. Last night he had not noticed a chit from the Adjutant that he wished to see him in the orderly room over the matter of some highly inaccurate timings in his last orderly officer's report.

Note: Due to gangs of rifle thieves from the North West frontier, all muskets were chained to sentries' waist belts, so that the Pathans could not stab a soldier and carry off his rifle. Buck shot was more deadly in the dark than a rifle. On training, rifles were buried in a pit, made under tent, and slept on by the occupants.

For your information –
In the scrap book of the 2/Royal Scots Fusiliers was a statement in his own writing by a sentry...... The battalion at that time was in Barrackpore near Calcutta, on to Hoogly. At 13 30 hrs. I saw a lion approach my post – I challenged it – it did not answer, I discharged one round of buck shot (the Dhobi's donkey was found wounded the next day – the soldier was one of the last draft).

Page 54

The Officers' Mess of an Indian Regiment on the line of march up the Grand Trunk Road

When an Indian regiment of cavalry or infantry changed their station, they usually went by line of march.

At about every 15 miles along the main roads there was a 'parao', or army camping ground. These were marked by white-washed pillars and had a deep well at a corner of the camp, well away from the road. Trees had been planted round the camp many years ago. Often inside the camp there would be a low wall enclosing the lonely grave of some British officer or soldier, who had died of fever or been killed in some 'affair' or 'skirmish' with Dacoits, rebels or Sikhs.

At the entrance to the camp, as the regiment rode or marched in, there would be a small party of old pensioners in spotless white clothing and medals, tears in their eyes, waiting to greet them.

With that courtesy and kindliness which Indians inherit, they would greet the colonel and all officers, British and Indian, and then wander round the lines talking to relations, NCO's they knew as recruits and the sons of old friends.

If a cavalry regiment, they viewed the horses with dim but knowing eyes.

After the officers had washed the dust of the plains off in their canvas baths and put on blue patrols they went to the mess tent where the pensioners would be waiting.

The officers then had the usual mess dinner except that the old soldiers insisted on paying for it. Most generous of hosts, they kept saying "Eat up sahib", "Have more to drink". As they sat round they also drank, except for the Mussalmans.

After dinner and several toasts, they joined the officers outside the tent as their guests.

Old days were discussed, an action on the frontier, what had happened to old Colonel 'Ismith' Sahib Bahadar who always sent us Christmas cards from his 'Klub' in London?

At last the colonel rose to say 'ijazzat', or permission to go. It would have been rude of the pensioners to ask to leave, as it would appear that they were bored by the company.

Leaving their presents of mangoes and eggs, they 'filed off' after much handshaking and promises to visit the regiment for the annual sports in the new station.

They went off down the narrow paths between the fields of corn and sugar cane, to their villages, under the vast star covered sky the only sounds those of the jackals and a dog barking far away, or the faint sound of the bugler sounding 'lights out', to dream of old days when soldiers of the regiment wore red coats.

Note: When the Seaforth Highlanders marched up the straight Grand Trunk Road to the Khyber Pass, the 'Jocks' used to cheer if they came to a slight curve in the road.

Page 99

Preparing a 'Kala Juggah', or Dark Place, for a Mess Ball.

It was most unusual for ladies to be allowed near a mess, but it was the custom to give them a great treat, at least once a year, and let them in.

It might be a dinner party at Christmas, or a ball to celebrate some famous battle in which the regiment gallantly destroyed thousands of foreigners.

The whole of cantonments looked forward to these events and 'dhurzees' were busy on verandahs with their sewing machines, churning out dresses copied from 'The Queen', 'The Lady' or an elderly 'Tatler'.

The mud floor of the dining room was covered by a drugget, well sprinkled with French chalk.

The colours were taken out of their cases.

A buffet was arranged in the ante-room, delicious eats from Karachi or Bombay were arranged round the mess silver.

The regimental band in full dress played foxtrots and waltzes with great feeling, due no doubt to ample supplies of free beer.

A lovely time was had by all, but romance was not forgotten.

'Kala Juggahs' or Dark Places, were carefully arranged by the mess servants on the verandahs for as 'Hobson-Jobson' stated in 1886 "The purpose of flirtation".

The sentimental servants looked forward to seeing some of the younger officers married off to the few spinsters in the station. New married officers meant employment for more servants and they had several relations in mind.

Two chairs were placed side by side very closely so that touching of bodies was impossible to avoid. Screens were placed round the chairs to ensure darkness and privacy from 'peeping toms'.

As Lady Dufferin wrote in 1885 'At night it was rather cold and the frequenters of the Kala Juggah were unable to enjoy it as much as I hoped they would'.

It is sad to relate that the Kala Juggahs were not usually occupied by the shy young officers they were intended for, but by older men in pursuit of an attractive wife whose husband was in the bar, dancing, or better still on a course 1,000 miles away.

It was a common occurrence for a screen and the ornamental pots of flowers to be knocked over when a respectable mem-sahiba feared she was about to suffer 'a fate worse than death'.

From left to right are – the Mess Sergeant, the Abdar who knows the drill backwards, the Kala Juggah, No.2 Khitmagar and the Head Mali.

The simple Mali is proud of his plant and his English learnt in a mission school. He explains that "precious flower comes out at night time smelling most sweetly and inspiring love thinkings".

Page 79

Bars in Messes

The scene is a station mess in India, at the end of the war. A Sikh Major who has long and gallant service in the Indian Army, is explaining the meaning of the expression 'Do Ungli Sikh Peg' (two finger Sikh peg) to two officers, whose knees are not yet brown. The forefinger and little finger are extended to measure a 'King size' whisky.

The barman is not amused, he is a Muslim, does not drink like Sikhs, or like them either. He has known better days with a famous regiment of Indian cavalry, when officers were 'pukka sahibs' relaxing in leather arm chairs after polo, pig sticking or shooting.

The colonel is slightly sad at the change, but supposed that he will be doing the same thing in that old inn in his village when he retires in a few months.

The new fangled idea of bars in officers' messes was introduced during the last war. These bars were usually made of three ply wood and painted in gay colours. They were placed in what had been the card room, or often in a corner of the ante-room. They were decorated with beer labels, advertisements for whisky and 'cut-outs' of ladies with few clothes on.

Officers even stood each other drinks, unheard of in the old days. Such expressions as "What's yours?", "Have this on me", "Let me buy you a drink" or "It's my turn to stand a round" were heard.

The older members who remembered more rigid days viewed the subject with sorrow, but kindness. In the East they felt it probably made the 'young fellows' feel more at home and reminded them of their 'local' and was better than them having drinking parties in their rooms.

In the past a bell was rung to summon the wine waiter. In India where bells rarely worked, a cry of "Koi-Hai" (anybody there) and an answer of "coming sahib" was the method of getting refreshment.

In officers' messes where British and Indian officers got on happily together, there was only one slight flaw. At meals there were separate dishes for the British and also the Indians.

The British officers far preferred the spicy curries to their own tough 'beef roast' and, as a result an Indian Officer arriving late discovered all Indian dishes 'were off the menu'!

Page 99

'Number One' brings in the 'Shroff's' bills on a guest night or 'The Shroff calls on the mess'.

It was the pleasant custom in the Far East for officers to sign chits for everything.

It might be a taxi to polo, drinks in a club, a champagne dinner at the best hotel in the company of some lovely girl of the 'top set' or bills in a Russian night club. In the latter the ladies were most attractive and charged one Chinese dollar a dance. Their father had always been in the 'Imperial Guard' and they quickly got thirsty after a few dances, and demanded champagne. This was some sort of tasteless gooseberry wine, but they got a 'rake off' and were sacked by the management if they did not ask for it.

If an officer was lonely, over fit and in need of comfort he went down to the Szechwan road where ladies of all nationalities were happy to entertain him. (Other ranks went to a meaner street close by.)

The officer signed a chit and went back to his barracks, as dawn broke, in a taxi and signed another chit.

He went on parade next day to watch his platoon do sectional rushes in some park. This tedious duty was relieved by an old Chinaman with performing mice.

The platoon, exhausted, also gathered round and enjoyed it as much as the officer. This was the only occasion on which he did not sign a chit. He fed the ducks and bears in the ornamental park with army biscuits. Later he played polo on his Mongolian ponies, he went to a tea dance at the Country Club or French Club, he signed chits, but never gave them another thought. He lived like a lord.

Alas, all chits were handed over by those owed money, to

Shroffs or debt collectors. These smiling gentlemen had little tact and would pursue an officer down the Bubbling Well Road as he walked with a rich Taipan's lovely, kind, intellectual daughter to a concert of classical music.

Bills for an enjoyable evening with Olga, Nina or Lotus Blossom a month before, were thrust under his nose. He was waved off with a smug smile. The Shroff usually turned up again at the officers' mess on a guest night when the Bishop was the guest of honour (he knew the officer had to be present).

Number One boy, or head butler, who had, like all the Chinese, a great sense of humour was only too happy to walk into the ante-room and say "Shroff say must catchee plenty dollars chop-chop" (for this he got 'cum shaw' but not as much as the Shroff).

Note:

Taipan – Number One of a firm (in India a rich Box Wallah).

Chop-Chop – pigeon English – make haste or look sharp. Supposed to be from Cantonese Kap-Kap or Mandarin Kip-Kip.

Cum-Shaw – Cantonese – Cam-Sau – a thank gift, that is a commission or squeeze.

Index